Food Safety, Risk Intelligence and Benchmarking

Food Safety, Risk Intelligence and Benchmarking

Sylvain Charlebois
Professor, Faculty of Management
Professor, Faculty of Agriculture
Dalhousie University

WILEY Blackwell

Registered Office
John Wiley & Sons, Ltd, The Atrium, Southern Gate, Chichester, West Sussex, PO19 8SQ, UK

Editorial Offices
9600 Garsington Road, Oxford, OX4 2DQ, UK
The Atrium, Southern Gate, Chichester, West Sussex, PO19 8SQ, UK
111 River Street, Hoboken, NJ 07030-5774, USA

For details of our global editorial offices, for customer services and for information about how to apply for permission to reuse the copyright material in this book please see our website at www.wiley.com/wiley-blackwell.

Library of Congress Cataloging-in-Publication data applied for:

ISBN: 9781119071129

A catalogue record for this book is available from the British Library.

Wiley also publishes its books in a variety of electronic formats. Some content that appears in print may not be available in electronic books.

Cover image:
Front Cover: tukkata/Shutterstock
06photo/Shutterstock
Fotokostic/Shutterstock
jimmyjamesbond/iStockphoto
Background: YUBO/Gettyimages

Set in 10/12pt Warnock by SPi Global, Pondicherry, India

Printed in Singapore by C.O.S. Printers Pte Ltd

10 9 8 7 6 5 4 3 2 1

Contents

Preface and Acknowledgment

The counsel of academics and experts from around the world was solicited to validate the metrics used in this study. These experts provided invaluable insight, leadership, and knowledge. I thank them for their contribution.

Most importantly, I dedicate this book to the people who allow me to pursue my passion, every day: My wife Janèle, my son Mathieu, and my daughters, Émilie, Judith and Laura.

1

Introduction

Facing Global Realities

This book includes the findings of 2010 and 2014 comparative studies of the food safety systems of some industrialized countries entitled "Food Safety Performance World Ranking." These reports were a follow-up on the 2008 inaugural world ranking. The purpose of this book is to place that ranking in a broader context through deeper analysis and a more expansive discussion of new, developing, and future issues in food safety unaddressed by the report.

Facing Global Realities

Global food systems connect all consumers. Food unites both global hemispheres through exchanges of commodities, knowledge, and technologies. The rich, the poor, farmers, and city dwellers, all are interconnected within food systems for the simple reason that everyone needs to eat. Many rarely think about it, but these links form a reality that is becoming more apparent as forces such as globalization and technology extend and intensify.

But while food systems bring us together, they do not always do so in positive ways. Increasingly, perceptions of fear and risk cause food systems from around the world to integrate (Spiekermann, 2009).

In the context of food safety, 2003 was a notorious year for Canada. During that year, the country diagnosed its first native mad cow case, and, in response, 35 countries issued embargos against Canadian beef. For the first time, a food safety-driven issue made headlines for weeks in our country. The United States did not allow Canadian beef in their country because of fear that it would compromise the value of their own beef products in lucrative markets like Japan (Lewis and Tyshenko, 2009).

Further major food recalls—related to spinach, peppers, and sliced meat—kept agriculture and the food industry in the public eye ever since.

Food Safety, Risk Intelligence and Benchmarking, First Edition. Sylvain Charlebois.

The Maple Leaf recall, triggered by a listeria outbreak that caused the death of 22 Canadians, is undoubtedly the biggest food safety story Canada has seen (Goveia, 2010).

In 2008, a global food crisis made media headlines and brought the topic of agriculture back to the front pages. Hunger, starvation, riots, and volatile civil unrest in numerous countries for several months occurred at the same time as record-breaking profits for Maple Leaf Foods (Pechlaner, 2010).

While the triggers of food crisis in 2008 were multifaceted and incorporated some environmental factors (such as climate change, droughts, and natural fires), many of these causes were human induced. These were structural and arose from societal decisions about the roles of agriculture, food, health, and regulation. Agricultural trade liberalization, the growth of biofuels, and a preference for commercial over subsistence agriculture in developing countries are a few instances of practices that influenced the crisis.

Food systems from around the world are exposed to mounting systemic pressures. In order to feed the planet, the world's agricultural output will need to increase by more than 40% in 2030 and by 70% in 2070 (Moeller, 2010). More than half of the world's population lives in an area with only a third of the world's arable land (Kelleher, 2010). The next decade is likely to see a major shift in global agricultural production and trade, and so system interconnectivity will become more significant though trades, exchanges, and strategic involvement.

The world has already shown that it can dramatically increase its food production capacity, but the situation today is different. Unlike at any other stage in history, water supplies are becoming scarcer and, therefore, irrigation technologies will be the key for agriculture. National governments are coping with shifting climate patterns that are challenging to predict and manage. We have recently experienced extreme climates that have affected crops and livestock producers from around the world. Responses and implemented policies vary from one country to another. In addition, interest in the environment and awareness of agriculture's carbon footprint is growing. Agriculture, which historically has been exempted from new environmental policies, is expected to undergo changes in years to come. Like other industries, agriculture will have to cope with environmental constraints that are both justifiable and a new challenge.

On the innovation front, genomics has played a significant part in augmenting our capacity to grow foods. This trend started many decades ago with arrivals of new genetically engineered crop seeds and will likely continue. Previously, the approach to agriculture was a linear thought process involving three Fs: food, feed, and fiber. However, methodologies such as genomics will soon change the relationships among these and other theoretical models.

Bioinformatics made it possible to sequence the human genome, thus enabling humanity to decode the basic instructions of life. Bioinformatics, or synthetic genomics, is recognizing the limitations of DNA management as DNA can break easily and becomes difficult to manipulate (Nicholson, 2009). The rise of bioinformatics has boosted the efforts of companies, most importantly in pharmaceuticals, to search for the right drugs and vaccines for particular diseases.

Bioinformatics will likely change our lives, but many wonder if food consumers and farmers are ready for these changes. We may be able 1 day to "print" mouse hearts, or even pig skin, literally (Beachy, 2010). But most consumers and farmers do not know what the term bioinformatics means, let alone how it will affect their daily lives.

Embracing biotechnologies can be a double-edged sword. It may not increase the risks to which consumers are exposed, but it will certainly alter those risks in many ways. Most importantly, the ways in which consumers perceive products crafted by new technologies will also change.

Agriculture's newfound prosperity, founded in part on growing connections with life sciences, is here to stay. The value of farmland around the world has increased significantly over the past decade (Bi et al., 2010). Farmers have been able to leverage their position and increase capital. Investments in many agricultural sectors are rising at an incredible rate. Agricultural technology and innovative farming methods are catching the attention of many farmers who have the financial means to invest. Of course, agriculture has always played a vital part in the economic development, but times are quickly changing. Food production may actually grow faster than anticipated.

The future ultimately relies on establishing a sustainable agricultural system and the exploration of alternative food solutions that will provide for all consumers. The global farming landscape has witnessed the arrival of new countries wanting to play a role on the worldwide stage. The path to a new world order is now on the horizon, yet there is no clear outcome. All we know for now is that, because of this influx of new wealth, the Western world has fewer but more efficient farms centering on the economies of scale.

Demand for food will also see its share of seismic shifts in the near future. The world's population will likely exceed 11 billion people by 2050 (Collins, 2010). It is estimated that over a billion people will reach the middle class by 2030 (Moeller, 2010). This will add a significant pressure to already-stressed grain supplies and fragment demand for available foods. Half the world's population suffers from some form of undernourishment from a scarcity of food, protein or micronutrients, or a combination of these (Schade and Pimentel, 2010).

China, India, and other emerging markets will greatly affect any food systems' capacity to address food security. In these countries, more and more peasants are fleeing the country for a better life in large urban centers and cities. In effect, more farmers are quitting food production and becoming consumers.

Urbanization is affecting lives, policies, and most importantly, the future of food systems. We have already witnessed this phenomenon in the Western world, but it is currently spreading around the world. Over the last decade, the world also has seen large migrations of people transferring across countries and continents. In response, food distribution systems must adapt.

Food Systems

If you consider all these factors, the ever-increasing complex exchange between food supply and demand has led to a greater focus on creating shared values between agriculture and consumers. One of these values is certainly food safety, and we will address that issue later.

So what are food systems? This book applies the systems approach (Hughes et al., 2008). Understanding the meaning of food systems is essential to appreciate their complexities. If we want to understand the entirety, parts of the food industry—such as production, wholesaling, retailing, or policy—cannot be analyzed in isolation.

By contrast, the food systems approach considers two basic and related components: elements of the food system and processes that make the system function. The elements of a food system are measurable things that can be linked together. For example, grocers can be linked to primary producers and domestic food-related policies can be linked to food-related policies found abroad. Everything is interconnected or interrelated. Food processes, on the other hand, change elements from one form to another (Morris and Reed, 2007). Food systems are comprised of elements and processes, a network which we call an ecosystem.

Systems can be open or closed. An open system is one in which external elements and processes alter its structure or functions. A closed system will always operate independently. Increasingly, it is argued that food systems are becoming more open than ever before. Food systems are open systems with respect to most elements and processes. They receive influences and inputs from their physical environment and, at the same time, cycle outputs back out of the system. They are also open to outside influences such as disturbances (e.g., embargoes, technology, trade agreements, *etc.*). Adopting a systems approach involves appreciating the scope and scale of the food industry, which is immense, complex, and difficult to simplify.

Food systems are being challenged more frequently due to the complexity of exchanges between elements, sometimes to the point at which the systems become compromised. Too often, global agrifood systems are characterized by the appearance of recurrent unwanted surprises.

In Canada, one such surprise occurred on May 20, 2003, when the country discovered its first native bovine spongiform encephalopathy (BSE) case, popularly known as mad cow disease. In response, many questioned the safety of our food chain. The Canadian Food Inspection Agency (CFIA) acknowledged that some meat from the infected farm may have in fact ended up on consumers' dinner tables (Anonymous, 2005).

At the time, the CFIA reassured the public that the likelihood of multiple cases among cattle of the same age is rare, and that the risk to humans of contracting Creutzfeldt–Jakob disease, the human variant of mad cow, is low. Unlike what British officials did in that country's mad cow crisis in the 1990s, when they tried to control consumer fears by concealing facts, the CFIA tackled Canada's mad cow scare by communicating the disease's real risks and by maintaining a science-based public dialogue.

However, the key to communicating intrinsic risks to consumers is not only to share scientific facts but also to manage systemic uncertainty. During the mad cow crisis, the CFIA showed its intolerance toward ambiguous situations, which it perceived as a threat, when it broadcasted to consumers information on the status of our food supply in the hope that information will keep a lid on ambiguity (Diekmeyer, 2008).

When people feel uncertain about the food they eat, trust is not a trivial issue. Regulatory officials can regain public trust only by offering protection and information that satisfies public uncertainty. Most observers agreed that government officials in Canada did not mislead the public during the BSE ordeal, even if uncontrollable variables hindered their capacity to predict the outcome of certain strategies (Nikiforuk, 2005).

So how real is the risk of contracting Creutzfeldt–Jakob disease for consumers? Even though Canada banned the practice of rendering ruminants for cattle feed in 1997, ruminant feed was still readily available on the market, and violations of the ban have been reported (Brooymans, 2005). Regulators found that enforcing the ban was challenging.

Another problem at that time was the CFIA's own assumptions about the disease. Some of the agency's leading veterinarians declared that animals younger than 30 months could not develop BSE. Japan, which has made BSE testing compulsory for all slaughtered animals, discovered two cases in 21- and 23-month-old animals (Kilman, 2005). Monitoring standards have since changed, which provide evidence that food systems do and are able to cope with changes in threats over time.

Within our food system, the CFIA walks a fine line between educating the public and trying not to alarm it, with the public's trust in the balance. Surveys over the years report that the vast majority of Canadian consumers unreservedly believe that our agricultural supply system is not endangering human health, and that they trust the safety of our food chain (Couture, 2009). But trust is fragile and can be obliterated in an instant. In other parts of the world, consumers were not so kind to food regulators and industries facing a BSE-driven predicament. By neglecting to nurture consumer confidence, industrialized nations such as Japan and Britain have paid a hefty price to regain the public trust their industries needed to regain profitability.

The Maple Leaf recall caused by a listeria outbreak in 2008 was another significant shock to our food system. Unlike the mad cow crisis in 2003, the Maple Leaf recall led to fatalities, 22 in total (Smith, 2010). Since the recall, the industry has changed. Certainly, Maple Leaf has changed: it revisited its protocols, and most industry elements were already following food safety practices that exceeded governmental regulations (Mason, 2009).

Some reports suggest that public authorities did not properly inform consumers about risks (Galloway, 2009). Consumers heard a confusion of voices and perspectives, which reduces the efficacy of every press conference, website, and article, as well as public investigations more generally. Shared accountability across supply chains should be at the forefront of any new food safety policies.

Occurrences such as the listeria outbreak at Maple Leaf made Canada, to an extent, food insecure. The recall had profound implications for Canadian consumers. As the Maple Leaf recall reveals, it is necessary that modern consumers understand that these episodes and their tragic outcomes can be minimized only by sound policies that address the complex, interlinked nature of our food economies.

Both events called for a systemic approach to food safety issues. Although they are very different, the Canadian mad cow crisis and the Maple Leaf recall are considered two pivotal incidents that changed how our food system operates. Since then, food safety became a common concern for most players within the food industry. These events, although they had negative consequences to consumers and organizations, depart from our society's previous expectations about how food systems should function. Complex, transnational issues like food safety, or other public health issues such as obesity, are major challenges that frustrate analysis and management by reductive methods.

Food Safety Systems

The foundations of food safety systems are similar to those of food systems generally. Safety—in the form of regulations, practices, and expectations—is conveyed from one element to another within the system. Exchanges allow information and resources to be shared. Supply chains must work in synchronization; participants are required to work simultaneously to provide safe foods to consumers. But relationships are bidirectional in nature. Systems calibrate through sharing responsibilities and become more accountable to one another. Consumers, too, are asked to share information with the system since they are intimately involved and part of it. The food chain across producer, processor, retailer, and consumer is highly interconnected and dynamic.

The chain of trust from suppliers to producers, to distributors, to wholesalers, to retailers, to end consumers, is essential for a highly functioning food safety system. A lack of legitimate representatives within the chain, failures to convince important stakeholders to participate, distances between participants, and the length and breadth of the supply chain are factors that limit joint action on crucial issues like food safety and traceability.

All elements play a key role, but consumers are our system's most central risk assessors (Labrecque et al., 2007). Consumers are the ones who risk, perhaps several times a day, buying food products from grocers, corner stores, street stands, eating at social events, and at other more or less familiar places. However, systems have demonstrated that they are often unable to provide information to the end consumer through proper traceability. Accordingly, it is noted that food choice is frequently swayed more by psychological analysis, such as perception of the brand, rather than physical properties of food products, such as the likelihood of food carrying a disease. Perception of food safety risk is skewed by psychological interpretations that influence attitudes and food buying patterns. Logic is habitually missing from consumer buying patterns. This fact can be explained as a result of the increasing incapacity of consumers to make their own assessments of the risk related to food threats and their dependence on public institutions to acquire strategic and suitable information (Markovina and Caputo, 2010).

More accurate assessments can be achieved through traceability. Traceability is an effective safety and quality monitoring system with the potential to enhance safety within food chains, as well as safeguarding the protection of consumers. Food traceability is the architecture behind all food safety systems. Shared responsibility throughout the food supply chain can in no way be evaded. Many have accepted that the BSE and Maple Leaf ordeals were part of a cycle in which conditions force us to enhance food safety systems already in place. Food traceability offers the ability to trace and track the origins of any product throughout the food supply chain, at any level.

When implementing a more universal traceability program, one has to keep in mind that food retailing is one of the most competitive industries in our global economy. Food retailers must manage disproportionate operational overhead costs, low profit margins, and demand that is relatively elastic for many products. Demand price sensitivity is the key when a food traceability project may increase retail prices. Moreover, on the other side of the marketplace, farmers are often considered price takers and depend heavily on governmental farm subsidies in order to survive.

Despite its costs, food traceability is a vital aim that reinforces accountability. For government and health officials, it means having the ability to act quickly in a crisis situation and know where animals or products are in the supply chain (Rosolen, 2010). By no means it can bulletproof the industry from major food recalls in the future, but it may permit anticipation of the these types of crises and adoption of proactive attitudes throughout the food supply chain, adding value to Canadian commodities in the process. It will also ensure more rapid containment, potentially in real time, of food catastrophes that could harm consumers.

The tools and techniques of food safety are related to the discipline of public health emergency preparedness: protecting and securing the population's health require information about food safety systems and consumers themselves. Like public health preparedness, food safety is heavily reliant on technology. The use of technology can leverage a food traceability system that may increase and improve the types of information elements that the system can share.

To consider an example, the world is calling out for nanotechnology, particularly in agriculture, where the technology could play a significant role. Nanotechnology offers the opportunity to manipulate matter at the smallest scale possible to date and allows engineering of functional food products at a molecular level.

Nanotechnology may lead to advances in agricultural research in the decades ahead. Applications of nanotechnology in agriculture and food systems include improvements to reproductive technology, conversion of agricultural and food wastes to energy and other useful by-products using enzymatic nano-bio-processing, disease prevention, and enhanced health of plants and animals. Researchers in Canada have developed nanofertilizers that release nutrients as plants need them (Moore, 2010).

Some predict that by 2020 the global impact of products in which nanotechnology plays a crucial role will be roughly $1 trillion (Canadian) per year with significant benefits to the food industry in food processing, ingredients, nutraceuticals, and delivery systems. Packaging will also benefit from nanotechnology, allowing for more efficient food safety monitoring (Dingman, 2008). Nonetheless, some have raised ethical concerns about nanotechnology and call for the contextualization of ethical discourse in its ontological, epistemic, and socioeconomic and political reflections (Ferrari, 2010). Open debate on nanotechnology is a prevalent topic among governments, research agencies, industry, and nongovernment organizations. With consumers, though, public perceptions about nanotechnology vary or are unclear.

Another noteworthy technology increasingly influencing food safety systems is radio frequency identification, also known as RFID. This technology has proven to be effective in traceability standards. The use of barcodes to identify products and lots has been the preferred technology since the late 1970s. Barcodes, however, are a read-only technology. RFID transceivers let data to be both read and written to a tag, which follows a product throughout the supply chain, providing stakeholders with better control and accuracy. Such a technology can increase the level of accountability for what is coming in and out of a facility. For the food industry, benefits from using RFID technology are higher reliability and higher rates of rejected products at the source.

More technologies in coming years will have a significant impact on how food systems assess, control, and contain food safety-related risks. It is difficult to pinpoint how far technology is capable of going and how willing consumers are to partake in sharing information within food systems.

The biggest challenge with large food recalls is finding the origin of the affected product, a task that rarely finishes as quickly as the companies and the public want. A food recall is often prompted by consumers registering at health clinics and hospitals after becoming ill from eating a contaminated product. The entire system must respond quickly upon recognizing these unofficial signals. Once recalled, hundreds, thousands, and sometimes millions of kilos of products are removed from the food chain, although the vast majority of it may not be unsafe. Proper technologies could trace and track products before and after process, throughout distribution, back to the processor, and even back to the farm from which a product came (Mehrjerdi, 2010). Accurate recalling is then more feasible. Existing technologies that allow this to occur are cost prohibitive for most companies and are ultimately thrown away by the end consumer. Research continues to supply affordable methods to the industry.

Supply Connecting with Demand

Food safety systems are influenced not only by supply-driven factors but also by demand, which is more fragmented than ever before. Elements of food systems are challenged by changing demographics, lifestyles, tastes, and attitudes. Most demand-focused trends are shaped by the rural–urban divide. The roots of the rural–urban divide lie in the historical strategies of centrally planned systems that favored industrial development and agricultural surplus largely for urban capital growth and urban-based subsidies. Residents of large cities represent well over 85% of most Western countries' populations. This trend has created a significant disconnection between agriculture and consumers. Many consumers remain unaware of how and where food is made and often take it for granted. Affordable food prices have further widened the rural–urban divide because consumers can buy more food while thinking about it less.

However, consumers are becoming increasingly conscious about the origins and health safety of the food they consume. This shift is not to be taken lightly. Local foods have enjoyed a resurgence in the past two decades in Canada, the United States, and elsewhere. Consumers are attracted to these markets by an array of environmental, social, and economic factors, often related to the alleged benefits of local food channels. Many consumers seek authenticity in reaction to the increasing intensity of marketing channels (Smithers and Joseph, 2010). The search for authenticity is a radical rejection of conventional, industrialized production methods and will significantly impact how food safety systems evolve. That being said, many consumers still adhere to industrialized systems and do not question their integrity.

The ideal food safety system aims at mitigating risks in real time. That of course, as stated earlier, would be an ideal. But given how public regulators are challenged by budgetary constraints, it is only for the longer term that one country should think of real-time food safety surveillance as being possible. Food safety is and will remain a challenge for all industrialized countries, but what will test food safety systems is the notion of food authenticity and fraud.

Food fraud is not a new phenomenon; historically, it dates as far back as the Greek and Roman Empire. However, in recent years, better access to advancing technology has allowed us to quickly recognize food distribution failures. As a result, contemporary food fraud has frequently found itself in the media spotlight. Currently, the authenticity of food in general and the veracity of food labels in particular are major concerns for many, including consumers, regulators, and the food industry at all levels of the food continuum. In light of the European horsemeat scandal, we have come to realize that failures such as mislabeling can occur at global scales. If you think Canada is immune to fraudulent food labels, think again. Chances are you have already unknowingly purchased a food product with an inaccurate food label.

Regulators and the food industry are beginning to realize that the problem is far more widespread than first supposed. For example, a recent study in the United States revealed a high substitution rate of 57% in meat labeling.

Consequently, there has been considerable dissonance between the contents of the product and the information found on its label. Similar results were found in a study involving chicken sausages in Italy. Not a month goes by now without a published study acknowledging how deeply problematic the situation is. This of course raises significant food safety and consumer protection issues. Allergens alone can pose significant risks to vulnerable consumers with medical conditions.

There are many reasons for the boom in fraudulent labels. The remarkable growth of food counterfeiting can be partially attributed to the increase in global trade, emerging new markets, and the steady increase in world food prices. Processors, agents, brokers, and distributors alike are often tempted to substitute ingredients or products to set an appropriate price point for a targeted market. In addition, resource scarcity, the potential for greater profits, and inadequate legislation have all encouraged, even made counterfeit labeling inevitable, the most common result of which is food fraud.

Over the years, some categories of food have been affected more than others. The most documented cases in the food industry have been with fish and seafood products, some of which have been reported in Canada. For example, DNA analysis of hake products commercialized in Southern Europe have demonstrated that more than 35% of fish packages were mislabeled on the basis of species substitution. However, in recent years, other categories have been targets: wines and olive oil, among others. There have been alleged cases of nonorganics being sold as organically grown commodities. The list, unfortunately, goes on.

The best solution for this problem is improved traceability. In the past, food traceability—the capacity to track food ingredients across supply chains—was promoted to improve food safety; it appears that increasing food fraud makes a case for the capacity for higher traceability. The enhancement of tracing systems alone is insufficient, however. Opportunistic behavior within our food system should also be monitored by food regulators. Unfortunately, the work of surveying the entire system regularly would be an overwhelming task. The Canadian food retail industry is a $120 billion business. It would be unrealistic and even undesirable to expect regulators to effectively monitor it. In addition, added public monitoring would likely result in increased bureaucracy and, certainly, higher food prices.

The food industry is just as concerned about food fraud as consumers; perhaps more so. Reported cases of inaccurate labeling can be devastating to both brand equity and the reputation of companies. The main driver for a reduction on food fraud cases is accountability within the industry, and consumers have every right to expect it. More questions are being asked when goods are sold from business to business before it reaches consumers.

For the industry, however, the clock is ticking. Technological advances are making traceability more accessible, and soon consumers themselves will be able to self-authenticate the origins of food products and the validity of ingredients. Many devices and apps to carry out this function are in development worldwide. It will be interesting to track the response of industry and regulators once consumers have access to these tools. Millions of citizen regulators may not be feasible today, but it is a very possible reality for the very near future.

The lesson is, before consumers actually become part of food traceability systems in real time, industry should ensure food fraud becomes a problem of the past, and as soon as possible. For that, food safety systems ought to think about the ways to include consumers in the process of risk mitigation, by running a much more open system, embracing a market-based approach. This is likely the only opportunity that both industry and governments have to enhance systems regularly. That would be a long-term goal. As food safety systems mature thought, it is important to recognize some important elements, which make regulators and the industry equally efficient. Metrics are presented in the next section.

Comparing Food Safety Systems

Food safety systems from around the world evolve at different paces. They vary from country to country, as do methods of risk assessment, standards, and policies.

The *Food Safety Performance World Ranking Initiative* was designed to facilitate identification of the relative strengths and weaknesses in Canada's food safety performance. The goal of this approach was to assist academics, practitioners, and policymakers in assessing food safety systems and processes in Canada. It allows for better risk intelligence by federal regulators around the world. Risk intelligence is a matter of adopting better risk assessment practices to monitor risks proactively.

This book is the evolutionary tale of international benchmarking in food safety performance. We have conducted three different surveys over the last few years in 2008, 2010, and 2014. We first compare results between the first two surveys in 2008 and 2010 (Chapters 2–6). After chairing a summit on regulatory food safety performance metrics in 2011 in

Helsinki, Finland, how we measure performances have changed. Another survey was conducted in 2014, and results are presented in Chapter 7 of this book.

Metrics were different for the first two surveys. They were established based on comments and recommendations made by a group of academics back in 2006 and 2007 during meetings held in Canada and Italy. For these surveys, in addition to measuring Canada's food safety performance, the report also investigated its underlying causes and highlights policies that could improve food safety in the future. This report compared Canada's performance with 16 peer countries across four major categories:

1) Consumer Affairs
2) Biosecurity
3) Governance and Recalls
4) Traceability and Management

The Consumer Affairs category measured policies and outcomes that assess how well countries connect with their consumers. Surveillance efforts, hygiene practices, and information accessibility are the main indicators.

The Biosecurity category concerned a country's capacity to contain all relevant risks related to food safety. This included the rate of agricultural chemical use and a country's bioterrorism strategy—the latter being an increasingly important aspect of food safety in the twenty-first century.

The Governance and Recalls category looked at the effectiveness of domestic regulations and governance related to food safety. For example, the existence of risk management plans, the level of clarity of food recall programs, and the number of food recalls were some of the metrics considered.

Finally, Traceability and Management measured a country's ability to identify the location of food items and its knowledge of a food item's history. This evaluation included the depth of the traceability programs.

Performance in each category was measured using only the indicators that reflect the overarching goal of the study. Eleven indicators were considered and evaluated.

The purpose of this benchmarking framework was to identify and evaluate common elements among global food safety systems. Therefore, the primary objective of this study was not only to identify which country offers the safest food products to its citizens but also to recognize which countries employ comparatively best practices to contain risks related to the safety of food systems. This study analyzes the performance of 17 top

Organisation for Economic Co-operation and Development (OECD) countries, including Canada. These countries are

Australia	France	Norway
Austria	Germany	Sweden
Belgium	Ireland	Switzerland
Canada	Italy	United Kingdom
Denmark	Japan	United States
Finland	Netherlands	

This group was used for all categories, variables, and analyses. Countries were awarded a grade of "superior," "average," or "poor" for each category.

Methodology for the First Two Surveys

As for the 2008 inaugural edition, the State–Pressure–Response model was used as the study framework. This is a useful approach to understand policy reactions related to food safety. The report considered outcomes that measure results—not effort. Indicators were divided into three classes based on the adaptation of the State–Pressure–Response approach used by the OECD to benchmark the environment. This model has three components:

1) State (output)—Condition of food safety performance at the time of the report
2) Pressure (input)—Human primary or secondary activities that impact the condition of food safety systems either positively or negatively
3) Response (policy and actions)—Policies and actions that the country has initiated or will initiate to address food safety issues

This study focuses on indicators that can be influenced by public policy. The factors that were taken into consideration are those that can be modified or altered by individual, organizational, or public efforts. Indicators may directly or indirectly influence output. For example, a policy that makes livestock identification mandatory may augment the capacity of a country to track meat products across the food chain, thus reducing foodborne illness.

All indicators used to measure performance within a specific category met the following criteria:

1) The indicator provides valuable information concerning the performance or status of the particular food safety domain.

2) The indicator can be affected by policy.
3) The indicator secondary data are reliable and readily available.
4) The data are sufficiently consistent to allow benchmarking over time and permit a valuable international comparative analysis.
5) There is general agreement that a change in the indicator in one direction is better than a movement in the other.

The data for this study were based on secondary sources, such as the OECD, the World Health Organization (WHO), the United Nations (UN), national statistical agencies, and other food safety regulatory organizations based on the countries under study. The most recent data were used for each indicator.

The choice of comparator countries was significant. This study compared Canada with other OECD countries because of the greater likelihood that these countries have achieved high standards in food safety. Initially all 30 OECD countries were to be considered, but some were later disqualified. Luxembourg and Iceland were dropped because both have populations of less than 1 million. In addition, the study only considers countries with a gross domestic product above the OECD mean. Therefore, the 11 countries that fell below this mean were also omitted.

The inclusion of emerging economies like India and China was a possibility; however, they performed poorly on food safety indicators. Furthermore, countries where food security is still a significant concern were not appropriate candidates either.

For output indicators, a ranking system of superior, average, and poor was adopted—comparable to a report card. Input indicators were not ranked because it is difficult to determine whether a higher value reflects higher levels in food safety performance. Moreover, it is difficult to establish any relationships between output and input. Response indicators used the same overall ranking system as output indicators. However, instead of using superior, average, and poor, grades of "progressive," "moderate," or "regressive" were used.

For the actual world ranking, countries were ranked for each category and results were then aggregated to generate a world ranking. As with response indicators, each country was given a grade varying between superior, average, and poor, thus creating three tiers.

Limitations

Some limitations ought to be considered. Secondary data was not always available for some of the countries studied. More specifically, the research on non-English countries (such as Japan) was challenging. Some countries are also less transparent than others, which makes the data collection process more intricate. A considerable amount of data were processed and analyzed with some level of subjectivity.

Secondly, this study includes research published between 2002 and 2010, which may skew results. Some agencies and countries publish reports every 2 years or so. In some cases, reports were only published once, which made it difficult to collect and consider current information on food safety.

Lastly, genetically engineered organisms (GEOs) are not considered in this study. When the project was originated, no conclusive evidence suggested that GEOs posed a health threat to consumers. Out-of-household consumption was also not considered because it would have made the variables more difficult to measure.

The evaluation of the 17 countries in this study occurs over four distinct categories comprising 12 criteria:

- Consumer Affairs:
  - Incidence of reported illness by foodborne pathogens
  - Rate of inspections and audits
  - Food safety education programs
  - Labeling and indications of allergens
  - Ease of access to public health information
- Biosecurity:
  - Rate of use of agricultural chemicals
  - Bioterrorism strategy
- Governance and Recalls:
  - Existence of risk management plans
  - Level of clarity and stability of food recall regulations
  - Number of protectionist measures against trading partners
  - Number of recalls
- Traceability and Management:
  - Depth of traceability system in food chain

Many scholars and practitioners from around the world reviewed these indicators, and data were collected and compiled for each category. Based on these data and the subsequent State–Pressure–Response model analysis, countries were ranked for each category, and results were then aggregated to generate a world ranking. Each country was given a grade of superior, average, or poor.

The world ranking and overall grade were derived in two ways. First, based on the grades over the four sections a country was placed in an overall grade category (superior, average, or poor). This informed a rough ranking, with superior-graded countries naturally ranking higher than poor-graded ones. Second, based on a country's average category rankings (between 1st and 17th), countries were then ordered within their overall grade category. For example, Belgium, France, Germany, Ireland, and Italy all earned overall grades of poor; however, Belgium's world rank is higher than Italy's because it has a higher category ranking average.

As illustrated by Table 1.1, Australia, Canada, Denmark, Japan, the United Kingdom of Great Britain and Northern Ireland (UK), and the United States of America (USA) all earned grades of superior, owing to their comparatively progressive category grades. Austria, Finland, the Netherlands, Norway, Sweden, and Switzerland earned average grades, owing to their overall moderate performance. Finally, Belgium, France, Germany, Ireland, and Italy all earned grades of poor for comparatively regressive category grades.

Table 1.2 provides the category-specific grades and ranks for the 17 countries.

Table 1.1 World ranking.

			2008 Comparison	
Rank	**Country**	**Grade**	**Grade**	**Rank**
1	Denmark	Superior	Superior	3
2	Australia	Superior	Superior	4
3	United Kingdom	Superior	Superior	1
4	Canada	Superior	Superior	5
4	United States	Superior	Average	7
6	Japan	Superior	Superior	2
7	Finland	Average	Average	6
8	Netherlands	Average	Average	12
9	Austria	Average	Average	14
10	Norway	Average	Average	9
11	Sweden	Average	Average	13
12	Switzerland	Average	Average	8
13	Belgium	Poor	Poor	16
14	Germany	Poor	Average	10
15	Ireland	Poor	Poor	17
16	France	Poor	Poor	15
16	Italy	Poor	Average	11

Increased grade	No change	Decreased grade	No data

Table 1.2 Category-specific grades and ranks.

Country	Average rank	Consumer affairs		Biosecurity		Governance and recalls		Traceability and management	
		Rank	Grade	Rank	Grade	Rank	Grade	Rank	Grade
Australia	6.25	5	Average	11	Average	2	Superior	7	Superior
Austria	9	16	Poor	5	Average	9	Average	6	Superior
Belgium	9.25	11	Poor	15	Poor	7	Average	4	Superior
Canada	7.25	2	Superior	11	Average	1	Superior	15	Poor
Denmark	3.5	3	Superior	1	Superior	8	Average	2	Superior
Finland	7.75	14	Poor	1	Superior	15	Poor	1	Superior
France	11.25	9	Average	14	Poor	13	Poor	9	Superior
Germany	10.5	13	Poor	5	Average	12	Poor	12	Superior
Ireland	11	4	Average	13	Poor	14	Poor	13	Superior
Italy	11.25	17	Poor	15	Poor	10	Average	3	Superior
Japan	7.5	8	Average	8	Average	4	Superior	10	Superior
Netherlands	8.75	6	Average	17	Poor	5	Average	7	Superior
Norway	9.25	10	Poor	3	Superior	10	Average	14	Average
Sweden	10.25	11	Poor	4	Superior	15	Poor	11	Superior
Switzerland	12	14	Poor	7	Average	15	Poor	*	N/A
United Kingdom	6.75	7	Average	9	Average	6	Average	5	Superior
United States	7.25	1	Superior	9	Average	3	Superior	16	Poor

Increased grade	No change	Decreased grade	No data

* Insufficient data.

The average rank column provides a category average rank for each country, which was used—inside of the overall grade category—to provide a number-based overall world ranking.

Comparison with 2008

There were few overall grade changes in 2010 compared with 2008, as only Germany, Italy, and the United States moved up or down a grade level. Moreover, each country earned a world ranking that was close to its 2008 result. Austria and Italy had the largest shifts (each moved five ranks), followed by Germany, Japan, the Netherlands, and Switzerland (each moved four ranks). Generally, the countries that moved the most did so in a downward direction: the largest shifts were negative (Italy, Germany, Japan, and Switzerland). This mirrors the overall grade shift, as the United States was the only country to increase its grade while Germany and Italy each fell one level.

In the 2010 study, each country received a similar grade and generally ranked close to its 2008 study result. This reflects two important aspects of this study. First, as a comparative study, a country decreasing in grade or rank does not necessarily mean that the country is providing poorer food safety systems or standards. Instead, this might simply mean that its performance is *comparatively* worse than its peers in 2010 compared with 2008. Second, some of the category variables (e.g., the measurement of the number of protectionist measures against trading partners) were measured using the same category standards as 2008. In these cases, more countries earned top-notch scores in 2010 than in 2008, which reflect an *absolute* improvement. In other words, the 2010 grades and rankings, because they are similar to the 2008 grades and rankings, reflect the speed in which some countries are improving their food safety systems—surpassing the changes in other slower countries.

Because this was a comparative study, the United States was able to increase its overall grade without increasing any of its criteria-specific grades. The United States simply scored comparatively better across all four categories in 2010, compared with 2008. By comparison, Germany and Italy both received lower grades because of their decline, compared with their peers, in the Biosecurity category.

Highlights

Generally, the non-European countries (Australia, Canada, Japan, and the United States) tend to perform fair equally. In part, this may be a result of a more-integrated agricultural and food safety system on the European continent. Moreover, with the exception of Japan, each of these countries have fairly large populations, land masses, and regional variations. It was not surprising that Japan also scored within this group as its unique food requirements ensure that it creates and adopts worldwide best practices.

Canada continued to perform very well compared with its international peers. Notably, Canada and the United States were tied in category grades, overall grade, and world ranking position. Like its continental neighbor and largest trading partner, Canada earned excellent grades and category rankings in Consumer Affairs and Governance and Recalls, performed decently in Biosecurity and fell short compared with the international average in Traceability and Management.

When Canada performed well (like in Consumer Affairs and Governance and Recalls), it ranked very well (earning second and first places, respectively). However, Canada's performance in areas where it is not setting international best practices was spotty, earning the country ranks of 11th and 15th in its average-graded Biosecurity and poor-graded Governance and Recalls, respectively.

As a result of increasing European integration, European Union (EU) member countries tend to perform similarly in many of the metrics. In part, this is a result of identical, EU-required national policies for several of the food safety criteria measured in this study.

The largest difference between EU-member countries emerges in areas where EU requirements are the weakest or individual national governments are required to demonstrate individual policy leadership. Perhaps ironically, the largest EU countries tend to perform the weakest compared with their smaller-state peers. This size distinction is most notable in the category grades, when individual State–Pressure–Response considerations are applied.

When applicable, Norway and Switzerland tend to perform best when adopting or participating in the EU or pan-European systems or standards. This specific pressure can be partly attributed to their participation in the common market.

Japan represents a unique food safety model, given its distinct needs. While it continues to perform well overall, Japan falls four world ranks to sixth in 2010. Special care was taken to ensure that the Japanese grades reflected its policies as they relate to the country's unique needs and pressures.

The categories with the largest grade improvements were Governance and Recalls and Traceability and Management. Biosecurity (owing in part to the new bioterrorism metric) was a generally poor category for all countries. It is particularly notable as the only category in which no country improved its grade compared with 2008. More data were available for the 2010 report, which helped to fill in certain gaps from 2008.

Results from the latest survey are intriguing. Some trends are emerging which may very well impact how food safety systems function and interact among themselves in years to come. In the next chapter, an in-depth analysis on how Canada's food safety system is performing is presented.

2

How Was Canada Doing in 2010? A Comparative Analysis

How Was Canada Doing? A Comparative Analysis

In recent years, we have witnessed the rise of food safety concerns in Canada. As food markets evolve, new technologies emerge, and as product differentiation and more affluent Canadian consumers increase, there is heightened focus on food safety and quality within our borders. Many have argued Canada is not doing well when coping with unwarranted threats to the efficiency of our food systems (Schmidt, 2010; Sweet et al., 2010). So much so that, in 2010, the Canadian government committed to strengthening Canada's food safety system. The government announced its intention to ensure that families are sufficiently informed to make healthy decisions, and that it will hold those who produce, import, and sell goods in Canada accountable for the safety of consumers (Kondro, 2010).

Nonetheless, by contrast with other developed countries, Canada is actually one of the best-performing countries. Its overall grade was superior, earning it a place among the top-tier countries, which average their category-by-category performance at the top of the scale. Like the United States (US), which it tied for fourth place, Canada earned two category grades of superior: one grade of average and one grade of poor. All other things remaining equal, if Canada were to increase its relative performance in Traceability and Management (from poor), it would be at the top of the world food ranking.

Canada's best category was Governance and Recall (ranked first overall) followed closely by Consumer Affairs (second overall). Its worst showing, as mentioned, was in Traceability and Management (ranked 15th overall). Canada performed in the middle of the pack on Biosecurity (11th overall). Overall, Canada's performance is very encouraging. Table 2.1 illustrates its grades and rankings across the four categories.

Food Safety, Risk Intelligence and Benchmarking, First Edition. Sylvain Charlebois.

Table 2.1 Canada's performance in all four categories: 2008 and 2010 grades.

			2008 Comparison	
Category	**Rank**	**Grade**	**Rank**	**Grade**
Consumer Affairs	2	Superior	3	Superior
Biosecurity	11	Average	14	Average
Governance and Recall	1	Superior	4	Superior
Traceability and Management	15	Poor	13	Average

As illustrated by Table 2.1, Canada improved its ranks in 3 out of 4 categories since 2008. It fell in Traceability and Management compared with its peers (from 13th to 15th), a drop that downgraded its mark from average to poor.

Highlights

- For a federal state, Canada is doing well at providing world-class and comprehensive food safety programs at the national level. Overall, it is ahead of its nearest federal state neighbor, the United States, in providing national food safety programs.
- Canada clearly articulates its programs, administration, and outcome data on its performance. This makes researching Canadian food safety much easier than it is in other, less forthcoming jurisdictions.
- Traceability and Management is Canada's weakest category. However, a new program is under development that could help raise Canada's category grade (and have a similarly positive impact on its overall grade and world ranking).
- Canada's lead food safety agency, the Canadian Food Inspection Agency (CFIA), compares very well with its foreign counterparts in the clarity of its regulations and processes. Its outcomes also score highly. Although the CFIA has been the subject of domestic concern stemming from the 2008 listeriosis (*Listeria*) outbreak, the Weatherhill report into that crisis noted that Canada scores well against its peer countries and that "for the most part, Canadians can have confidence in Canada's food safety system." Furthermore, the 2010 Food Safety Performance World Ranking study found that the CFIA scored highly against its international peers in some categories and poorly in others. This study's findings echo the Weatherhill report: "By [making improvements] Canada can raise its global food safety ranking from superior to the best in the world" (Weatherhill, 2009).

Canada's performance, especially compared with the United States, is very important to its agricultural industry. Good food safety programs assure Canada's trading partners that its food exports are safe. This is especially important for a trading country like Canada, which in 2009 exported about $35.1 billion (Canadian) in agrifood products worldwide (Agriculture and Agri-Food Canada, 2010) and (Canadian) 22.3 billion to the United States in 2008 (Agriculture and Agri-Food Canada, 2009).

Consumer Affairs

Consumer Affairs was Canada's second-best category in overall performance. While Canada scored a grade of average in its number of incidences of reported illness by foodborne pathogens (output), it scored a grade of progressive in each of the four policies.

Canada had a middling finish in the number of incidences of reported illness by foodborne pathogens but held its grade from 2008. It earned an average rank of seventh, comprising a third-place average in *Campylobacter*, a sixth-place average in *Salmonella* and *Yersiniosis*, but a worrisome 13th-place average for *Escherichia coli* (Similar to 2008, this study did not consider *Listeria* in its list of foodborne pathogens.)

Rates of inspections and audits were a better metric for Canada, as it was one of only three countries to receive a progressive grade. Likewise, it ranked very highly in food safety education programs and the ease of access to public health information.

Another Consumer Affairs highlight was Canada's increase to a progressive grade in its labeling and indications of allergens regulations. Thanks to strict product-labeling requirements, Canada was 1 of 13 countries to increase its grade from 2008 but only 1 of 3 to increase that grade to progressive.

Biosecurity

A weaker category than Consumer Affairs, Canada's Biosecurity earned a grade of average. The country uses a comparatively low level of pesticide per hectare of agricultural land, and its progressive grade is a large improvement over its 2008 performance. However, Canada did not perform as well in the new bioterrorism strategy metric. While it does have a bioterrorism strategy, this policy was only good enough to earn the country a grade of moderate.

Governance and Recall

Governance and Recall was Canada's best comparative performance. It earned the highest grade and ranking (superior and first place), largely due to the CFIA. In each of the individual metrics, the CFIA (which is responsible for Canada's risk management, food recall regulations, and food recalls) performs very well compared with its international peers.

The CFIA administers a rigorous, clear, and science-based risk management system. It also maintains a clear and consistent food recall regulatory program, which has met and continues to meet all of the CFIA's recall–notification performance targets, including the *Listeria* crisis in 2008. In addition, the CFIA oversees about 350 food recalls each year, which in 2008 included 192 during the *Listeria* crisis alone. The 2009 *Report of the Independent Investigator into the 2008 Listeriosis Outbreak* found a number of normative problems with the CFIA (Weatherhill, 2009), including issues with the CFIA's risk management, food recall regulatory program, and communications. However, the 2010 Food Safety Performance World Ranking considers Canadian policies—not outcomes—as they compare with international peers. Therefore, some of these normative suggestions, while potentially important, were not issues at work in the rankings.

Canada earned only an average grade for its number of protectionist measures against its World Trade Organization (WTO) trading partners. The country has a somewhat low, but not ideal, agricultural most favored nation (MFN) rate. Nevertheless, compared with its peers in this 17-country study, Canada performed very well in the Governance and Recalls category.

Traceability and Management

Traceability and Management was Canada's weakest comparative performance: it scored a grade of poor and was ranked last (because Switzerland earned a grade of "Not Applicable"). While the CFIA earned Canada top marks in Governance and Administration, its traceability system is regressive compared with those of its peer countries. The CFIA system does not cover live animals, and is therefore not comprehensive (covering all food products), nor does it consider food along the entire farm-to-fork continuum. Currently, the CFIA and partnering provincial and industry participants are developing a farm-to-fork system. While this system is not yet launched, an improvement in this category could raise Canada's Traceability and Management category grade and have a large impact on the country's overall grade and world ranking.

How Canada Got Here

The regulation of food safety, governance, and settings of standards can be traced back to the Roman Empire. In the thirteenth century, consumer protection became a more prominent concern for Western European civilizations as agricultural trades mounted among nations (Ginsberg, 2007). Weights and measures, food purity, and false product labeling were focal issues at the time. We saw nations becoming more concerned with unsafe commercial practices, leading to the implementation of food safety regulations. The nineteenth century marked the emergence of chemical contaminations. In order to better understand these threats, new disciplines such as food science and chemistry materialized. By the end of the nineteenth century, new technologies which allow for better food preservation and conservation were developed. That thrust was consistent with the industrial revolution and the increased demand for processed food around the world.

Throughout history, farming and agrifood industry sectors serve as the basis for socioeconomic development, capitalist market transformation, and the integration of nations into the global market system. More recently, national governments face multiplying food safety risks due to unclear and deficient regulation. Specifically, in the United Kingdom, the introduction of foot-and-mouth disease and bovine spongiform encephalopathy (BSE) triggered new standards that affected many countries, including Canada. To counter these looming threats, countries created collective structures to govern food safety in global agrifood industries.

However, many attempts to harmonize regulations within and across nations have failed. Such an approach requires a normative consensus, policy cooperation, and legal convergence to protect consumer health and the public interest. Food safety regulation now attempts to address the systemic risks posed by global consolidation and vertical integration of the agrifood sector. As such, governments increasingly rely on public interest models of regulation based on precaution and science-based risk assessment.

In North America, the ratification of the North American Free Trade Agreement (NAFTA) in 1994 bolstered the agrifood and biotechnology sector throughout the continent (Bello, 2008). The Canada–US Free Trade Agreement (CAFTA), enacted in 1989, established technical standards and regulations in the dairy, poultry, eggs, cotton, sugar, and sugar products sectors. But it was NAFTA that greatly propelled food and agriculture trade within North America. The United States, Canada, and Mexico established (if not exactly an equally tripartite agreement) at least a stated intent to improve resource allocation. This would be accomplished by the removal of barriers to cross-border trade in goods and services; increased cross-border investments with included legal protections, promotion of fair competition, and equal treatment among competitors in all three countries; and creation of institutions to implement agreements and resolve disputes.

All parties of the NAFTA agreement have benefited from increases in trade flows as a result of this venture. Between 1989 and the 2008 financial crisis, exports to the United States had more than quadrupled from about $60 billion (US) a year to $280 billion (US) a year. Almost 25% of those exports are in the food and agricultural sector (Cuddehe, 2009). In addition, US food and agriculture trade with Canada and Mexico accounted for 28% of exports and 35% of imports in 2008. Before NAFTA came into effect, Canada and Mexico accounted for only 17% of all US food and agriculture exports and 25% of US agriculture imports (Maniam et al., 2003).

Essentially, the NAFTA agreement largely drove the integration of the North American food processing industry. Trades in biotechnologies and an increase in direct investment to expand agrifood processing facilities leveraged economies of scale and other industrial efficiencies. Consequentially, food costs were lowered significantly, product quality was standardized, and consumer choices converged within the three NAFTA countries.

The North American context certainly has influence over Canadian agrifood policy. Food safety is one of the main pillars of Canada's Growing Forward Agricultural Policy Framework (APF), started in 2009 to replace the 2003 APF. One of the framework's objectives is to make Canada the world leader in food safety. Like its predecessor, the framework is the result of collaboration among federal/provincial/territorial governments. The new framework builds on the APF by offering a more holistic and strategic approach to agriculture and delivers improved programs and services for Canada's broad agricultural sector. The framework allows ongoing initiatives to ensure that both levels of government and the agrifood industry work in concert to provide consumers with a high measure of protection (Jarrett and Kobayakawa, 2008). Its focuses on the following areas: building a competitive and innovative agriculture and supporting a sector that contributes to society's priorities and that is proactive in managing systemic risks. Contrary to the first generation of the framework in which food safety was treated as a separate category, food safety is enmeshed in all of the three main pillars or the new framework (Bednarek and Ahn, 2010).

Canada's high dependence on export markets plays a significant part in the country's aims for its standards to be harmonized with those of other nations. Canada is a net exporter of food and agricultural products, notably agricultural grains and livestock. Wheat is the most traded commodity for Canada: over 20% of traded wheat around the world is Canadian. Other Canadian-traded commodities representing between 2 and 4% of global product are coarse grains, oilseed, pork, beef, veal, and poultry. Still, Canada is also dependent upon imports of certain food products (Gray, 2008). The cost of food imported by Canada, largely fruit and vegetables, fish, coffee, and tea, had fallen in 2010 mostly because of the rising exchange rate of the dollar and lower prices on world markets for many foodstuffs other than grains. Despite the fact that Canada is endowed with many resources and affordable land, Canada's share of world agrifood trade is only around 3% (Bow, 2010). Nonetheless, Canada's dependence on exports provided an incentive for the country to discourage other countries from issuing nontariff barriers. Thus, aligning safety standards, regulations, and inspection systems with the United States is crucial to Canada since 60% of the country's agrifood exports and imports are with Americans.

Domestically, intergovernmental relations play a crucial part in developing food safety standards in Canada. With two major levels of government involved, achieving a high-quality food safety system is complex (Frampton, 2008). Responsibilities are not always shared, but accountability, in the eyes of the public, is perceived as similar for both levels of government. In practice, the federal government oversees quality and grading standards for trade interprovincially and internationally. The federal government's role also includes preventing the domestic commercialization of dangerous food products. By contrast, provinces and municipalities handle provincially based commercial activity. Municipalities invest their energy on food retail outlets, such as restaurants, food stores, and temporary and mobile food outlets.

If we look closely at the federal level, duties and responsibilities are actually divided between two organizations. Both Health Canada and Agriculture Canada are responsible for food safety in Canada in different capacities. Health Canada's responsibilities relate to products that have yet to enter the food system. It also monitors products derived from biotechnology and administers food labeling.

Health Canada is also known to study topical issues that can be of significant concern to policymakers and industry.

Health Canada has tackled sugar, trans-fatty acids, and sodium in recent years. In 2007, a working group recommended reducing sodium and that Canadians should reduce their daily consumption of sodium by almost 35% to 2300 mg before 2016 (Eade, 2010). While this is not an "ambitious" target and critics decry the voluntary nature of the recommendations rather than a prescriptive requirement to lower sodium content in foods, this report is interesting because it attempts to make "sweeping changes to policies, behaviours, perceptions and other daily decisions" (Charlebois, 2010). Importantly, consumers, industry, government, researchers, and nongovernment organizations were all represented, which reflects Health Canada's recognition that the complexities of food systems require a global perspective, food industry acumen, and a realistic approach (Charlebois, 2010). Unlike the 2005 trans-fatty acid labeling requirement, the committee took a more nuanced role, recognizing the role that sodium plays in our diets and the reality that "salt adds flavour and may be good for you, when consumed in moderation" (Charlebois, 2010). (Interestingly, after 2005, the "trans-fat-free" label became a selling feature for food companies even before policies were enforced (Martin, 2009), which is unlikely to happen to the same extent with sodium.)

The other federal organization that plays a significant role within the Canadian food safety system is the CFIA. The CFIA's basic function is to enforce food safety standards across the country. The CFIA conducts border inspections for foreign pests and diseases to prevent the introduction and spread of quarantined pests which could impact Canada's grains and field crops sectors. Although Canada has a voluntary-based food recall system, the CFIA may issue recalls and enforce regulations related to fraudulent labeling incidents.

The CFIA and its significance to the Canadian food safety system were certainly challenged during the *Listeria* outbreak at Maple Leaf Foods in 2008. The episode has solidified in Canadian consumers' minds as one of the worst foodborne illness outbreaks in the country, and the CFIA deserves some blame for this. A side effect of this notoriety is a disconnect between consumers who believe that Maple Leaf took the right actions to alleviate the crisis and those skeptical about the safety of the company's products.

Although Maple Leaf forms a large spot on the public radar, the CFIA's role in delivering information to consumers during the crisis barely registered. Both organizations need to be more accountable to consumers if they are to recover from the product recall and become productive players in a trusted Canadian food industry.

In a recent survey by the University of Regina, 95% of respondents remember hearing about the Maple Leaf product recall more than 6 months after 20 people in Canada died from listeriosis, and several others were sent to hospital (Couture, 2009).

Even though 65% thought that Maple Leaf Foods managed the recall well, 40% of all respondents and 25% of those identified as customers of the company said they have not eaten Maple Leaf products since the recall. These figures are worrisome, considering the good press Maple Leaf Foods received in the aftermath of the recall. This lack of trust translates into lost sales for the company and diminished brand equity.

Yet consumers believe that Maple Leaf is managing the situation well and rightly so. The August 2008 mega-recall of products has transformed the company, particularly its food safety processes and protocols. Production executives at several levels are now compelled to follow strict standards that go beyond regulations. Regardless, two in every five consumers remain unwilling to eat Maple Leaf products and as many as half of them believe the products to be less safe than they were before the recall. There is a disconnect between belief and trust when it comes to modern food safety confidence.

Trust is a crucial indicator of an organization's effective response to a crisis. It is also a central channel through which brands are constructed. But trust is inherently fragile. It is slowly gained and may be rapidly lost. Once lost, trust may require a long time to rebuild, if it ever does.

While the survey does not claim that Maple Leaf Foods has forfeited the trust of consumers, it suggests that the outbreak dented that trust. What is reassuring for the company is that consumers did not feel "betrayed" by it, because they perceived that Maple Leaf demonstrated sound leadership throughout the ordeal. Company chief executive officer Michael McCain and his staff were considered credible and accountable. These are assets required to build a constructive dialogue among consumers, the CFIA, and other corporations in the food industry to promote decision-making processes that can safeguard Maple Leaf's brand equity. To grow, the company must recognize that consumers are still concerned about the safety of its products and thus continue taking innovative actions to allow consumers to engage in a direct, transparent, and sustainable relationship with Maple Leaf Foods. One such method was the YouTube video, posted in August 2008, where McCain explained what Maple Leaf was doing to contain the epidemic and apologize for the consequences of the outbreak. Maple Leaf Foods should continue to be open to dealing directly with consumers and not only through food distributors, retailers, or restaurants. The CFIA's failure to relay information to the public may be a big factor in Maple Leaf's current situation.

The survey gathered data on the sources from which consumers were accessing information during and after the recall. The most important sources, not surprisingly, were television and newspapers. Only 2% of consumers reported that they consulted the CFIA for information. One probable issue is that the agency hotline or website is not as readily available as newspapers or broadcast news reports. The CFIA should take note. Its shortcomings in disseminating information during crises should be addressed with a more proactive role in jargon-free communication with consumers about risks. Part of the reason for low consumer trust in Maple Leaf products may be the perceived lack of information derived from a trusted governmental body that could coordinate and reinforce food safety decisions. The CFIA is not doing its job if it is merely content to distribute statistics. Instead, the agency should work independently and in concert with companies such as Maple Leaf to communicate with consumers to help bridge the gap between belief in a company's efforts and trust in its products. In cases where there is a food safety concern, the CFIA needs to communicate with Canadians in a clear, effective, and proactive way. Similarly, when a concern has been addressed, the CFIA must make the new *status quo* clear to consumers. In cases where the food safety issue concerns one company in particular, given the CFIA's inspection role, the agency also has an obligation to maintain clear, effective, and proactive communication with consumers about the status and safety of a particular producer, including when the problem has been resolved.

Industry and the Canadian Government

Best practices in food safety and risk management evolve over time. For example, Hazard Analysis and Critical Control Point (HACCP) has become a prominent strategy for Canadian industry. As a matter of fact, in Canada and the United States, HACCP systems are mandatory for federally registered processing plants in a number of sectors, including meat, fish, and seafood (Hobbs, 2010). Other countries have adopted HACCP as well. The most important driver of HACCP deployment is customer requirements that HACCP to be implemented in supplier facilities. Under the CFIA's direction and supervision, the food industry follows strict practices prescribed by HAACP. The US influence over food safety standards exposed Canada to a considerable need to adopt HACCP practices. Consistency with wider international standards is another factor. HACCP is recognized under CODEX as an internationally accepted standard for food safety. The CODEX Alimentarius is a collection of standards, recognized by the WTO, which covers all types of foods (Vogel, 2010).

To some extent, industry is already ahead of government in areas such as improving food safety and environmental conduct beyond legal minimums, guidance of governments, and trade rules. These efforts respond to specific consumer concerns as well as potential liability limitations, over and above governmental regulation. The industry has a significant role to play in linking food, human health, and the environment. A paradigmatic divide exists in Canada between the public and private sector and their respective missions to protect the public. Around the world, governments tend to apply a moralistic approach to risk containment, while private stakeholders are concerned that new regulations would impact the regular flow of business (Sapp et al, 2009). On the one hand, public officials are under increasing moral, legal, and political requirements to protect consumers and public health from unsafe food and food safety risks in food systems. Risk communication and management strategies are often challenged by abrupt disturbances, which may escalate into food safety crises. Also, concerns over the economic impacts of food safety crises on national food producers and industries somewhat delay the enforcement of risk management and food safety measures. Both governments and the private sector often engage in a balancing act between risk mitigation and managing fear, and Canada's organizations are no exception. They share the same objective of protecting the public but often have divergent interests. Pressures from government, industry, and consumers influence stakeholders' food safety outcomes and decisions.

The food industry often perceives new regulations as trade-diverting or as outright harmful to trades. There is evidence that private standards can be trade enhancing, diverting, or reducing under different circumstances, and the outcome can be expected to differ across products or industries. The food industry is highly complex. The competitive architecture of the food retailing sector and the degree to which private standards are chiefly proprietary or broadly adopted consensus standards are some of the key determinants of the trade impact of private standards (Hobbs, 2010).

Tensions between industry and the Canadian government were never so obvious than in 2003, when Canada experienced its own mad cow crisis. The Canadian cattle industry, major victim of the crisis, has a distinctive history. That history impeded the abilities of the cattle industry and government to cope with the crisis (Labrecque and Charlebois, 2006).

The end of the American Civil War in 1865 brought food shortages to the aboriginal peoples of the North American plains. The bison herds upon which they had depended were being eradicated. To help meet the demand for meat, the United States contracted cattle producers to push large herds of Texas longhorn cattle north toward Western Canada. This marked the beginning of the Canadian cattle industry. The large regions of grazing land attracted foreign investment, and the Western Prairies were rapidly occupied.

As part of that process, Prairies opened up to homesteading. Most farmers owned only a few head of cattle and horses, kept primarily for work and basic needs. Energy and money went into the production of wheat rather than beef. By the end of the 1930s, tractor power began to replace animal power. In the years that followed, this and other technological advances resulted in the increased availability of feed grains, particularly barley. Beef cattle became an important component of mixed grain farms, and Canadian cattle numbers in the West increased from three to nine million between 1940 and 1975 (Agriculture and Agri-Food Canada, 2002).

During the 1950s, the use of corn silage enabled Central and Eastern Canadian producers to finish cattle more economically than their Western counterparts, whose cattle were still being finished on the range. Improved economic conditions and the ready supply of Western calves for finishing enabled a large feedlot industry to develop in Eastern Canada.

Climate, availability of coarse feed grains, and improved marketing and transportation alternatives fostered a substantial feedlot industry in the early 1970s. Today, the Canadian beef industry is an integral part of the Canadian economic mosaic.

Canada is known as a country of agricultural production surpluses. The Canadian beef industry, which generated $7 billion (Canadian) in revenues in 2002, has always been perceived as producing high-quality commodities on global markets. In 2002, beef and cattle imports in Canada were valued at $1 billion (Canadian), whereas exports of beef and beef-related products to all countries were estimated at $4 billion (Canadian), almost 85% of which were exported to the United States. This makes the industry predominantly dependent on its international trading partners, particularly the United States and Japan.

On January 30, 2003, a 6-year-old Angus cow in the Canadian province of Alberta, sent to slaughter at a provincially licensed meat packer (provincially licensed packers cannot export their products), was initially diagnosed as having pneumonia and was put down before entering the food chain. Unfortunately, it was not until May 16, 2003, that the sample was tested and found positive for BSE. The diagnosis was confirmed again by the CFIA and at Weybridge veterinary laboratory in the United Kingdom (Duchesne, 2003). On May 20, the CFIA announced its first-ever native BSE case to the world, thus igniting an industry-wide crisis. Exports of Canadian beef and cattle were immediately disrupted. Nontariff trade barriers were enacted across the world (Canadian Press, 2003a). Most importantly, the United States shut down its borders to Canadian beef. Within hours, many other countries, including Japan, Mexico and Thailand, followed suit. The CFIA immediately started its investigation. It destroyed and tested 2700 cattle in Western Canada (Canadian Press, 2003b). Although no other cases of BSE were found, the Canadian beef industry had lost access to major markets.

May 20, 2003 is considered to be the clinching event of the Canadian mad cow crisis, equivalent to the "Black Wednesday" of the British mad cow crisis of 1997. Some scholars consider it the founding act (Pauchant and Mitroff, 1992). A founding act triggers a crisis, and signals a total breakdown in the collective sense-making of a marketing channel (Pearson and Clair, 1998). However, the first Canadian domestic case of BSE was detected in a British-born cow in 1993, 3 years before the 1996 British report that linked BSE to vCJD, but drew very little public attention.

Since then, food safety concerns have influenced economic and political policies employed by regulative institutions around the world. Additionally, most countries have opened up their markets to increase trade with their international partners. Political and economic alliances between countries in Europe, North and South America, and Asia, as well as the creation of the WTO, have remodeled the premises of international commercialization (Buzby, 2003).

After the first native case of BSE was diagnosed in Canada, sentiments of helplessness and distress exacerbated trade disturbances, uncertainty, and power disequilibria in the beef-marketing channel. Several beef producers blamed food manufacturers and distributors for refusing to stimulate beef demand, through techniques such as decreasing retail prices of beef products offered to consumers on the domestic market. In addition, many observers have argued that financial compensations from public funds given to beef producers have been disproportionate. Unquestionably, many political clashes and setbacks among partners within the marketing channel have transpired since the crisis started.

This industry-wide crisis was initiated by embargoes issued against Canadian beef-related products by many of Canada's foremost international trading partners. Their decisions transformed interorganizational relationships within the Canadian beef-marketing channel. More specifically, these circumstances created uncertainty that hampered the decision-making capacities of channel members within the Canadian beef industry. Observable variables from the Canadian BSE crisis led us to believe that economic and sociopolitical forces are capable of redefining power and dependence-relation conditions within food marketing channels. Recent international trade agreements and global food safety concerns have propagated external political forces. The influx of agreements and trading regulatory agencies has utterly changed the geopolitical symmetries between nations. Before the Canadian BSE crisis, food insufficiency weakened the power of countries that were not able to produce an abundance of food supplies. Nowadays, international trading agreements enhance interdependency among nations and have made supplying countries, like Canada, rather dependent on foreign markets to absorb excess commodity surpluses and food products. When embargoes were issued against Canadian beef across the world, its selling price dwindled and disturbed the Canadian beef industry's inner sociopolitical structure, creating still more uncertainty for channel members.

During the Canadian BSE crisis, channel members were desperate to resolve the issue as quickly and thoroughly as possible. The patterns of behavior of several key agencies related to the Canadian beef industry were predictable. Many channel members settled for ineffective defense mechanisms. Managers often appeared unable to deal with the emotional, informational, and cognitive aspects of the crisis events. By refusing to consider their own vulnerability, uncertainty was increased throughout the marketing channel, reducing the possibility of establishing any long-term organizational and logistical changes to the industry. As a result, the Canadian beef industry has been the focus of intense scrutiny by both public and private institutions in bids to reassure consumers and restore lost markets. Knowing that events similar to the BSE crisis may reoccur, many institutions in the industry wonder how an industry-wide crisis can be appropriately managed in the future.

At first, many channel members, including regulatory institutions, stated that the BSE-infected cow was an isolated incident, and that the media had inflated the entire affair (Canadian Press, 2003c, 2003d). Meanwhile, producers asked several levels of government for financial compensation, politicians blamed other agencies and jurisdictions, and many other actions were taken by channel members to immediately improve their strategic situations without considering long-term implications (Canadian Press, 2003e, 2003f).

Industry and the United States

Beyond investors, economic downturns do generate more casualties. The usual suspects, climate change, social programs are likely to suffer. But when the debt ceiling is raised in the United States, food safety quickly becomes a target, as it is in Canada. Because of the deal, the Congress has had difficult decisions to make as they worked to trim almost a trillion dollars from a wide variety of programs, including the FDAs and USDAs, key food safety regulators in the United States. The enforcement of the Food Modernization Act, which President Obama signed into law in 2012, has been compromised as will the inspection mandates.

In January 2011, food safety was very much a priority in the United States. President Obama signed a $1.4 billion overhaul of the nation's food safety system with the Food Safety Modernization Act, the first major food reform since 1938. The Food Modernization Act was meant to become an audacious attempt to provide Federal Agencies with more authoritative power to proactively intervene in the event of a looming outbreak. Canada, which has a voluntary-based recall system, would have had likely no choice but to make more policy changes related to food safety regulations, primarily to please our most important trading partners. Deepening financial turmoil in the United States, and Europe for that matter will put the implementation of new food safety measures on the back burner.

Paradoxically, all of this came on the cusp of one of the largest food recalls in US history, which garnered very little media attention in Canada. Agribusiness giant Cargill just issued a recall of £36 million of ground turkey in what's being called the largest food recall ever.

Nevertheless, food safety agencies in the United States are facing major cuts which will likely exceed $300 million.

These circumstances remind us that food systems cannot solely rely on governments to protect consumers from outbreaks. Economic cycles and political tug-of-wars, as preposterous as they seem to be, compel even the strongest willed governments like the Obama administration to consider food safety as a secondary issue. Surely, the Obama administration would never admit to that but money talks.

One in six Americans who can work are out of a job and many are living in a house worth less than what is owed on them. In some parts of the United States, food security is becoming a much more important issue, let alone food safety. That is why food stamps are immune to looming cuts.

The industry will have to occupy more space and play a more proactive role in the implementation phase of the Act, and they can afford it. The same is for Canada. Despite major economic headwinds our global economy is facing, second quarter earnings season was strong, and the mood surrounding earnings remained remarkably upbeat, up until early August that is. Most food companies are doing well financially, so bearing the food safety burden is very much possible. For example, the Act encompasses a food traceability pilot project which was to stretch over 18 months. Food traceability systems allow food supply chains to better trace and track products in case of a food recall. These systems are usually quite onerous and very expensive to implement, but it seems as though the food industry has more means than it ever had in the last decade or so. Around the world, including in Canada, industries are usually ahead of the food safety regulation game, but trades and consumer trends are making food safety compliance more challenging. At times though, some do miss the mark. The scope of food recalls, like the one at Cargill, reminds us how complicated food systems have become.

A better partnership between food safety regulators and industry has merit and should prevail, despite the fact that governments around the world would not be able to invest further in food safety systems, at least for the next little while. Canada is not immune and the CFIA has faced the same reality. The industry now has yet another opportunity to demonstrate some leadership and become even more accountable toward the public, while governments are slowly licking their financial wounds. That will likely continue for quite some time.

Beyond BSE: Food Safety and Trades

BSE and other food safety crises in Canada have shown that much of our food safety policymaking, whether explicit or implicit, lacks a cohesive direction. Certainly, trade policies have influenced public policies on food safety, but the science has developed faster than policies or the managerial capacity of national regulators to oversee food safety measures, which makes any BSE crisis a sociotechnological disaster (Denis, 1993). Most countries would base their food safety decisions on a risk analysis approach, but approaches can vary. This risk analysis basis may explain why many countries issued embargoes on Canadian beef despite the amount of scientific evidence showing that the product was safe to eat (Phillips, 2001).

Food policymaking is essentially a sociopolitical process but extends beyond that sphere. Most industries are in a productionist paradigm, focusing mainly on output and trades, and fail to synchronize production and consumption (Lang and Heasman, 2004). Many current agricultural public policies around the world concur with this paradigm. Most of all, food safety policies and regulatory systems heavily depend on the food system and the private sector for information, advancements in technology, and sharing and processing of data. The private sector has shown some degree of initiative by adopting safety systems such as the HACCP. As well, most nations now rely on other nations for food variety and food supply. The dynamics of the industry as a whole have utterly changed in such ways that increase the likelihood of future crises.

On the international scene, there are significant differences among nations and governmental authorities in approaches to food safety policies, particularly in terms of trading livestock between Canada, the United States, and Japan. In many countries, jurists and lobby groups now control the politically charged food safety arena, and there are historical reasons for this. In the 1980s and 1990s, world governments were significantly less interventionist with respect to public policies on food supplies, and instead let the market determine the directions of change and distribution. These practices quickly shifted when global food safety concerns arose. The governmental structures of many countries were not prepared for these changes. Regulators are often enmeshed in conflicts of interest or are perceived to have dual roles. For instance, in the United States, the federal Department of Agriculture's unwillingness to change food safety policies derives from conflicting mandates: on the one hand, they are tasked to provide safe and quality foods to the American consumer; on the other hand, their job is to promote consumption and marketing of American-made commodities. Canada has engendered a similar predicament. Traditionally, the department related to agriculture has held these dual responsibilities, even though the safety mandate would arguably better fit the overall assignment of a health department. Expanded information, shared accountability, and cost involvement are issues that have triggered many debates within food safety and supply chains.

For these and other reasons, food safety issues have become a source of conflict between governmental departments and supply chain members within and between countries. Around the world, food safety is a multifaceted and political issue, and many countries are adopting protectionist measures in order to cope with market uncertainty. Science and risk management practices are sadly less consequential than the whims of policymakers.

Politics is, and always has been, an integral part of food safety policy, and the methods used by governmental authorities and industry to cope with the BSE crisis is a sign that an adjustment of strategic paradigms would be beneficial. Beneath the politics and conflict of international food safety, there is, in theory, one simple solution among others to safeguard our food chain and minimize risks for our foreign customers: a transversal food traceability system that will track Canadian food products from the producer to the consumer, from their origin to our plate. As we have seen, traceability is one of Canada's weaknesses. Unfortunately, a practical application of this solution involves some significant problems. In government or industry, this initiative has often been conceived in a sectoral manner, whereas what is required now is the integration of sectoral interests in our policy framework for an efficient food traceability system. The costs are difficult to evaluate, and, for agrifood businesses, the lack of longitudinal vision has caused systemic inconsistencies. Consequently, agrifood businesses exposed to this collective project have concluded that they do not have sufficient financial resources to support such an endeavor. Most agribusinesses agree that the government should assume all financial and social burdens arising from collective projects. However, regardless of who pays for it, the implementation of a rigorous traceability system has become a fundamental need, and the capacity of the beef industry to adjust to these new realities is an incontrovertible requirement following from the adoption of new technologies.

Government and industrial authorities must find ways to modify the structure of the beef industry in order to facilitate the implementation of an efficient food safety framework and food traceability system. Without such a system, the Canadian beef industry remains vulnerable to the politics and lobbying of the international food safety arena, as would other countries in similar situations. To establish this system within the Canadian beef industry, certain paradigms must change. This new approach should introduce national branding strategies and specifically focus on food quality and country of origin labeling. The tactical efforts of the last 10 years will need a universal, strategic, and inclusive agenda that combines all forward-thinking paradigms of the industry into one. The productionist paradigm that currently overrules all other approaches is already in conflict with public health, and will quickly become obsolete.

For any given nation, food strategists will have to accept that domestic and foreign food safety policies are slowly converging. This does not necessarily mean that all standards across nations will become identical. It is very unlikely that the world will ever apply homogenous food safety standards, as food safety policymaking is a politically charged process. Food marketing strategists will have to consider the most rigorous of standards as being the ideal toward which they should strive. Food traceability systems and standards will have to comply with this new global reality, and it is up to food strategists and policymakers to achieve these goals.

As well, the North American legislative dynamics in which Canada has to operate is somewhat different than those observed in Europe. Only three countries are part of the North American political landscape, one of which is still considered by many nations as the world's only superpower. Because of trade ambiguity and distortion, standardization and normalization of food safety policies across countries are often governed by the most powerful entity. Observations made during the Canadian BSE crisis suggest that the United States determines the food safety policy schema for the North American continent. Conversely, the European Union, though dwarfed by the United States, commands significant economic power, and many of its constituent national economies are growing. Yet, France, Germany, and the United Kingdom are forced to compromise due to economical countervailing, a situation somewhat analogous to relations among Canada's provinces.

Canada's food safety system has faced its fair share of tests in recent years. Overall, Canada's performance is nearly exemplary. But food safety performances are indeterminate given the unpredictable nature of systemic risks. Therefore, Canada should not rest on its laurels. We currently have a two-tiered approach to food safety, which entails both provincial and federal standards that—together—reduce Canada's ability to streamline food safety processes across supply chains. In fact, food safety regulations are such a burden already that it is arguably more difficult to start a processing plant in Canada than it is to start a hospital.

More inspectors and more regulation may cost more without any proven benefits for taxpayers and consumers. When BSE struck Britain in 1997 and Japan in 2001, citizens of those countries saw food retail prices increase alongside legislation and regulation. Unless consumers are willing to pay a premium for food safety that is disproportionate to the risk, we need to be careful what we wish for. There is no evidence that more inspectors will make our food safer. France, Belgium, and Denmark have increased their rate of inspections without significantly reducing foodborne illness outbreaks.

The real problem is that our food safety system stokes consumer fears through the obsession provincial and federal regulators have with scientific data that is dangled in front of a concerned public without being placed in proper context. The CFIA's mission statement is "to excel as a science based regulator, trusted and respected by Canadians and the international community." Rigor and verifiability are definitive principles in any scientific method. In risk management, nurturing a scientifically driven culture makes sense. But a purely science-based approach does not give consumers a complete picture and can lead to the mistaken view that spending vastly larger sums of money on inspection will vastly reduce the risk.

Food distribution and the logistical complexities of food safety systems are abstract concepts. The real battlefield is in consumers' minds—in risk perception rather than real risk. Perception is critical and no one is doing an appropriate job of managing it properly. Canada does not necessarily need more regulation. The problem lies in the architecture of the system itself. With our current resources, we may be able to handle two significant changes. First, the CFIA may need to alter its dual mandate of protecting the public and assessing risk. By dividing these duties, Canada could establish an independent food safety agency that focuses solely on consumer concerns. Second, we may need to reconceptualize our food chain in its continental context and develop an approach to food safety that does not increase obstacles to international trade. Europe, Australia, and New Zealand have adopted similar approaches with great success. Since our economy is highly integrated with the United States, we owe it to our consumers to engage in a dialogue with US authorities.

Food safety is about consumer confidence and not just risks. Regulators and legislators are concerned only with safety and risk, not with perception. There is no evidence thus far that Canadian consumers distrust the safety of our food supplies or that there is widespread cause to do so. But trust is precarious and we need the proper food safety architecture to maintain it. Responsible regulation helps consumers recognize risk and take appropriate steps to manage it. If consumers succumb to panic, however, and unwisely assume more inspectors is the answer, what they will buy with higher grocery bills is the delusion that more regulation equals less risk.

Let us see how other countries are managing risks. The next chapter will begin with an investigation on Consumer Affairs.

3

Consumer Affairs

Connecting with the Consumer

The *Food Safety Performance World Ranking Initiative*'s investigation into Consumer Affairs considers one outcome (number of reported illness by foodborne pathogens) and four policies (rates of inspections, food safety education, labeling and indications of allergens, and access to public health information). Countries were ranked based on their performance in each of the five criteria on a sliding scale from superior to poor.

In evaluating the five criteria, country-specific judgments were made to smooth out some of the missing data. For example, Italy received grades of not applicable in two of the five criteria (it received two regressive and one moderate in the others). Therefore, Italy's actual grades for the two not applicable categories were assumed to be similar.

As illustrated by Table 3.1, Canada, Denmark, and the United States (US) ranked highest in Consumer Affairs, each earning a grade of superior. These countries performed well overall, with the United States attaining high grades in each of the five criteria. Australia, France, Ireland, Japan, the Netherlands, and the United Kingdom (UK) achieved a grade of average, with variable or overall midlevel performance. Austria, Belgium, Finland, Germany, Italy, Norway, Sweden, and Switzerland performed poorly, with several regressive grades and large gaps in their data.

Comparison with 2008

Trends over the last few years across the categories show some interesting and perhaps unexpected results. Denmark and France each improved their grades compared with 2008, both raising their scores in three of five criteria. Denmark raised its rate of inspection grade, France improved its foodborne illness numbers, and both increased their grades in two categories: labeling and indications of allergens and access to public health information.

Food Safety, Risk Intelligence and Benchmarking, First Edition. Sylvain Charlebois.

Table 3.1 Ranking for Consumer Affairs.

			2008 Comparison	
Rank	**Country**	**Grade**	**Rank**	**Grade**
1	United States	Superior	2	Superior
2	Canada	Superior	3	Superior
3	Denmark	Superior	8	Average
4	Ireland	Average	9	Average
5	Australia	Average	7	Average
6	Netherlands	Average	1	Superior
7	United Kingdom	Average	4	Superior
8	Japan	Average	6	Average
9	France	Average	14	Poor
10	Norway	Poor	5	Average
11	Belgium	Poor	12	Poor
11	Sweden	Poor	16	Poor
13	Germany	Poor	15	Poor
14	Finland	Poor	13	Poor
14	Switzerland	Poor	10	Average
16	Austria	Poor	11	Poor
17	Italy	Poor	17	Poor

Increased grade	No change	Decreased grade	No data

Norway, the Netherlands, Switzerland, and the United Kingdom fell in the Consumer Affairs category in 2010. The Netherlands dropped its inspection, food safety education programs, and provision of public health information grades; Japan's ranking fell because of labeling and allergen information concerns; and Switzerland declined as a result of (comparatively) increased foodborne illnesses and poorer provision of public health information.

Analysis of Ranking Data

Incidences of Reported Illness by Foodborne Pathogens

This first Consumer Affairs metric analyzed five common foodborne pathogens: *Campylobacter* (*Campylobacter jejuni*), *E. coli* (*Escherichia coli* O157:H7), *Salmonella* (salmonellosis), *Vibrio*, and *Yersinosis* (*Yersinia*).

This comparison illustrated the country-by-country rates of illness by presenting the number of cases per 100 000 people (to correct for population differences) over a 10-year period (2000–2009) and interprets this trend as an upward or downward variable.

The interpretation of the data can be done in many different ways. The most important limitations of this comparison are the quality of the data and how they are analyzed and interpreted. Several considerations are important to put the results of this study in perspective. Based on the available data, it is difficult to speculate on reasons why variations in the monitoring and surveillance systems between countries are so significant. Borders seem to matter in some cases. However, the study suggests that some countries either apply less surveillance or their system does not allow for more reporting to occur. It can also be possible that levels of pathogen reporting are affected by cultural differences in how risk is perceived from a broader sense. The study opted to reward countries with favorable trends which suggest more rigor in how regulators survey risks. This choice was not trivial as food systems have matured over time. The essence of this phenomenon is captured in more details in the last chapter of this book.

It should be noted that there are some inherent limitations of these data. This would apply to all the data presented in this book. In this case, variations in reporting practices and interpretation of disease definitions result in qualitative and quantitative biases that affect the representativeness and completeness of the data set. In the case of Canada, data were compiled from the Canadian Notifiable Disease Surveillance System (CNDSS) and the Notifiable Diseases Reporting System (NDRS). Whenever there were no available data, the National Enteric Surveillance Program (NESP) data were used. It is important to note that because the isolates reported in NESP is only a subset of laboratory isolations, they may not reflect the actual incidence of disease. For the purpose of this report, the highest value reported from these agencies was used.

In the case of the United States, Campylobacteriosis and Yersinosis are not nationally notifiable conditions. Active laboratory and population-based surveillance are conducted in FoodNet sites, which was used as the source of data for this report. Despite the fact that FoodNet population is comparable to the US population, the findings might not be extrapolated to the entire population of the United States.

The 10-year average per 100 000 people was calculated for each country by taking the published rate of illness in each country by year and dividing the sum of those rates by the number of annual data reports available. Each country's average then received a positive or negative shift according to the difference between their last recorded rate and that average. That last calculation evaluated whether a country is currently above or below its 10-year average.

For example, Australia has a 10-year average of 39.12 cases of *Salmonella* per 100 000 people. However, its 2009 rate was 43.6 cases per 100 000 people—a 4.48 case per 100 000 people increase over the 10-year average. As a result, Australia's total *Salmonella* score was 43.6 (the 10-year average of 39.12 plus the 4.48 difference between the 2009 rate and the 10-year average rate).

Each country also received a pathogen-specific ranking (e.g., Australia ranked 11th for *Salmonella*). A country's comparative scores were then averaged into a cumulative ranking. The top four countries received grades of superior, the next four received a grade of average, and the rest received poor grades.

France, the United States, the Netherlands, and Ireland all received superior grades, each ranking among the highest in several pathogen types compared with their peers. With the exception of the Netherlands, each received at least one top finish ranking for a specific pathogen. Nevertheless, Ireland and the United States both ranked quite poorly in *E. coli*. Austria, Norway, Canada, and Belgium all received average grades. Austria and Belgium received decent ranks for all of the pathogens, while Norway and Canada ranked highly in some categories and lower in others. The United Kingdom, Australia, Finland, Switzerland, Germany, Denmark, and Sweden received grades of poor. These countries uniformly ranked at or near the bottom for every foodborne pathogen.

There was no available foodborne pathogen information published for Japan. As well, Italy has not released sufficient data to include it in this metric.

Comparison with 2008

As illustrated in Table 3.2, there were two new additions to the superior grade category, as France and the United States joined Ireland and the Netherlands. Both of these new additions ranked the highest in 2010, with France ranking the highest in two of three pathogen categories and taking third place in the third.

Denmark, Finland, and Switzerland fell into the poor-graded category, scoring well into the bottom percentiles of several of the pathogen categories. Belgium, which scored a poor grade in 2008, rose just enough to enter the average-graded category by scoring just well enough in each category to increase its grade. Unlike two of its average-graded peers, Canada and Norway, Belgium was consistently in the middle of the pack in each pathogen category.

Canada held its grade from 2008, falling in the middle of the average-graded category. It had an average rank of seventh, comprising third place in *Campylobacter*, sixth place in *Salmonella* and *Yersinosis*, and a worrisome 13th for *E. coli*.

Table 3.2 Incidences of reported illness by foodborne pathogen: 2008 and 2010 grades.

Country	2008 Grade	2010 Grade
Australia	Poor	Poor
Austria	Average	Average
Belgium	Poor	Average
Canada	Average	Average
Denmark	Average	Poor
Finland	Average	Poor
France	Average	Superior
Germany	Poor	Poor
Ireland	Superior	Superior
Italy	Poor	N/A
Japan	Poor	N/A
Netherlands	Superior	Superior
Norway	Average	Average
Sweden	Poor	Poor
Switzerland	Average	Poor
United Kingdom	Poor	Poor
United States	Average	Superior

Increased grade	No change	Decreased grade	No data

Japan and Italy were removed from the grading system in 2010 owing to insufficient data. At least five years of data, per pathogen, were required to facilitate the calculation of a 10-year average. Because of the variable nature of foodborne illness outbreaks, evaluating a country on too few years of data would introduce new variables into the equation. For example, with only a few years of data a country might score particularly high or particularly low because of too few or too many outbreaks in a single year, compared with a proper average. Sufficient Italian data are unavailable for *Campylobacter*, *Yersinosis*, and *Vibrio*. Data are available for *Salmonella*, for which Italy ranked first. However, ranking Italy first overall, based on one data point, would have been inappropriate.

There is no published data on Japan's rates of foodborne pathogens. However, the academic consensus was that foodborne pathogens are significant risks to Japanese food safety. Part of this risk arises from cultural dietary practices. For example, "in Japan where seafood is eaten raw,

70 per cent of food-borne human illness is seafood associated" (Ferri, 2005). The primary cause of this illness is *Vibrio*. According to a United States Department of Agriculture (USDA) special report, "[m]icrobial food-borne illnesses pose a significant health problem in Japan" (Genthner, 2010). In addition to *Vibrio*, large outbreaks in Japan include *E. coli*, *Campylobacter*, and *Salmonella*. Nevertheless, without year-by-year outbreak data, it was impossible to compare Japan's rate of food-borne illness with its peer countries. Therefore, like Italy, Japan also received a grade of not applicable.

Rates of Inspections and Audits

Rates of inspections and audits were measured by two factors: whether a country has a national inspection policy that requires a minimum number of inspections and the number of inspections and audits carried out under this mandate. This metric looks at the level of effort regulators provide to better food safety systems.

Canada, Japan, and the United States all received grades of progressive for having strict inspection policies and carrying out many inspections. A recent comparison between Canada and the United States revealed a discrepancy between the two countries' inspection policies with Canada requiring only weekly inspections of meat packaging plants while the United States requires daily inspections (Galloway, 2010). Belgium, Denmark, Germany, Ireland, and the Netherlands received grades of moderate for their limited or assessed-risk-based inspection policies. Finally, Australia, Austria, Finland, Norway, Switzerland, and the United Kingdom received regressive grades for limited or subnational inspection policies.

There was no data to grade the inspection and audit policies of France, Italy, and Sweden.

Comparison with 2008

As illustrated in Table 3.3, Denmark was the only country to increase its grade from 2008 to 2010. Like many of its neighbors, Denmark conducts risk assessment-based inspections: focusing its inspection efforts *where the need is greatest* on a targeted, rather than constant, basis. However, Denmark also has a universal *Smileys* system of public grades for its food producers and resellers. The very public, universal, and rigorously inspection-based grades for local food safety earned Denmark an increase to moderate.

Like Denmark, many European countries use risk-based assessments and targeted inspections. The other moderate countries—Belgium, Germany, Ireland, and the Netherlands—all monitor and inspect foods

Table 3.3 Rates of inspections and audits: 2008 and 2010 grades.

Country	2008 Grade	2010 Grade
Australia	Moderate	Regressive
Austria	Regressive	Regressive
Belgium	Moderate	Moderate
Canada	Progressive	Progressive
Denmark	Regressive	Moderate
Finland	N/A	Regressive
France	N/A	N/A
Germany	N/A	Moderate
Ireland	Moderate	Moderate
Italy	N/A	N/A
Japan	Progressive	Progressive
Netherlands	Progressive	Moderate
Norway	N/A	Regressive
Sweden	N/A	N/A
Switzerland	N/A	Regressive
United Kingdom	N/A	Regressive
United States	Progressive	Progressive

Increased grade	No change	Decreased grade	No data

for safety but are not as directly involved or do not monitor as diligently as the progressive-graded category countries. In Belgium, the Federal Agency for the Safety of the Food Chain is responsible for all inspections, and employed 775 people to carry out 64565 sampling missions and 106610 inspections in 2008. However, these inspections were based on a risk assessment rather than a universal policy of inspection and audit (Belgian Federal Agency for the Safety of the Food Chain, 2009). In Germany, the Federal Office of Consumer Protection and Food Safety undertakes risk-assessment-based monitoring of producers, while also testing an annual basket of goods for foodborne health risks. Ireland also conducts risk-based inspections by prioritizing its food producers. In 2008, 82% of high-risk-categorized plants were inspected, 70% of medium-risk plants, and only 29% of low-risk plants (Food Safety Authority of Ireland, 2009). The Netherlands operates its inspection policy by "supervising the supervisor." In this case, the supervisors are farmers, growers, and fishermen, who are required to make daily checks and inspections of their own products and procedures.

Australia, Austria, Finland, Norway, Switzerland, and the United Kingdom all received regressive grades because either they do not have a clear national food inspection policy or the national government does not conduct inspections. Australia currently relies on a system of state and territorial policies. The Food Standards Australia New Zealand (FSANZ) organization, a binational food safety regulator for Australia and New Zealand, is currently developing a national code for food safety standards for Australia. National standards already exist for seafood and dairy, with egg, poultry, raw milk products, seed sprouts, and meat products currently under consideration. In Austria, the number, type, and methods of food inspection are difficult to pin down. The Austrian Agency for Health and Food Safety (AGES) is responsible for inspections. However, it claims that "the field of food inspection reacts quickly to new problem areas," but then states that it follows and evaluates problems identified by the EU Rapid Alert System for Food and Feed (RASFF). It then hails the "efficiency of the food inspection aptitude of AGES" as evidenced by the lack of German and UK "food contamination scandals" in Austria. Finland received its regressive grade by not having a national system of meat inspectors, though it does have plant and feed inspections—"food control in practice is largely the responsibility of municipal veterinarians and health inspectors" (Finland Ministry of Agriculture and Forestry, 2006). Norway, similarly, does not set its inspection policies at the national level and prefers to adopt and use EU policies. While the Norwegian Food Safety Authority (NFSA) does set regulations and conduct inspections, there is little information available on the agency's independence from the European Union. Switzerland does set its own federal policies on food safety (Switzerland Federal Office of Public Health, 2009); however, inspection duties are carried out by the subnational Swiss canton authorities. In fact, the national government only intervenes or controls food safety operations, through the cantons, in the event of an emergency (Switzerland Federal Veterinary Office, 2008). Like Switzerland, the United Kingdom conducts food inspections at the local level, with policy set at the national level.

Canada, one of three countries to receive a progressive grade, operates a hybrid system of inspections and policy. The Canadian Food Inspection Agency (CFIA) "[collaborates] with public health authorities at the federal, provincial and municipal levels" (CFIA, 2006). Similarly, the US Office of Food Safety inside the United States Department of Agriculture (USDA) regulates and conducts inspections, in cooperation with state and local authorities (USDA, 2010). Japan, which also has a multileveled system, splits responsibility among its national government, the prefectures and local authorities. This system includes 300 federal inspectors and over 13 000 local, nationally regulated inspectors (Japan Ministry of Health, Labour and Welfare, 2010). Unlike Canada and the United States, Japan cannot feed its population without imported food (Japan Ministry of Health, Labour and Welfare, 2010). This results in an import-skewed focus. For example, between April and September 2009, Japan inspected 108 390 import items comprising 11 791 tons (Japan Department of Food Safety, 2010).

Food Safety Education Programs

Food safety education programs were measured according to their number and scope. Unlike other metrics in this category, education programs are considered as a preventive measure. It allows us to better understand how countries tend to preemptively anticipate risks by empowering consumers with more information on how to protect themselves. Countries with several multifaceted programs targeting all or nearly all populations scored the highest. Countries with some information or limited programs scored moderately well. Finally, countries with no food safety education programs earned regressive grades.

Canada, Ireland, the United Kingdom, and the United States all scored grades of progressive for having large, multifaceted programs with near-universal coverage. Australia, Denmark, France, and Sweden have significant programs; however, they are not as comprehensive or universal as the progressive-graded countries. Finland and Switzerland scored regressive grades because of their extremely sparse programs and information. Similarly, Austria, Belgium, Germany, Italy, Japan, the Netherlands, and Norway do not publish data; therefore, they too earned regressive grades.

Comparison with 2008

As illustrated in Table 3.4, Australia and Ireland increased their 2010 grades. Australia provides significant information on its website, while Ireland boosts its food safety resources even further with more specifically targeted programs and resources.

Ireland and the United Kingdom, with their comprehensive, state-run food safety education programs, compare favorably with the other European countries included in this study. Denmark, France, and Sweden provide limited or nonuniversal programs; however, their domestic food safety agencies do provide and publicize some information. The other European countries, Austria, Belgium, Finland, Germany, Italy, the Netherlands, Norway, and Switzerland, provide no, or next to no, information for the public.

Table 3.4 Food safety education programs: 2008 and 2010 grades.

Country	2008 Grade	2010 Grade
Australia	Regressive	Moderate
Austria	N/A	Regressive
Belgium	Regressive	Regressive
Canada	Progressive	Progressive
Denmark	Moderate	Moderate
Finland	Moderate	Regressive
France	Moderate	Moderate
Germany	Moderate	Regressive
Ireland	Moderate	Progressive
Italy	Regressive	Regressive
Japan	N/A	Regressive
Netherlands	Moderate	Regressive
Norway	N/A	Regressive
Sweden	Moderate	Moderate
Switzerland	N/A	Regressive
United Kingdom	Progressive	Progressive
United States	Progressive	Progressive

Increased grade	No change	Decreased grade	No data

Japan is alone among the non-European countries under study in providing no information to the public.

Canada and the United States provide excellent, comprehensive, and cooperative food safety education campaigns at the federal and subnational levels of government. These programs earned these federal states progressive grades.

Labeling and Indications of Allergens

A country's labeling and indication of allergens grade considered the stringency of its mandatory regulations for food products across a variety of information points. The comparative metrics were name of item, use by or made on date, nutritional information, list of ingredients, country of origin, warning of allergens, and indication of food additives.

Australia, Canada, and the United States earned grades of progressive for their strict and detailed requirements for labeling. Most importantly, all three require specific allergen information. Austria, Belgium, Denmark, Finland, France, Germany, Ireland, Italy, the Netherlands, Norway, Sweden, and the United Kingdom (the EU countries, plus Norway) received grades of moderate. Their regulations require most of the elements for a progressive grade; however, they are notably missing allergen warnings.

Japan is the only country to earn a grade of regressive. Its food labeling regulations require the least amount of information compared with its 16 peer countries. There was no information on Switzerland's food labeling and allergen information regulations.

Comparison with 2008

As illustrated by Table 3.5, Australia, Canada, and the United States increased their grade from moderate in 2008 to progressive in 2010. With the exception of the United States, each requires all of the labeling

Table 3.5 Labeling and indications of allergens: 2008 and 2010 grades.

Country	2008 Grade	2010 Grade
Australia	Moderate	Progressive
Austria	Regressive	Moderate
Belgium	Regressive	Moderate
Canada	Moderate	Progressive
Denmark	Regressive	Moderate
Finland	Regressive	Moderate
France	Regressive	Moderate
Germany	Regressive	Moderate
Ireland	Moderate	Moderate
Italy	Regressive	Moderate
Japan	Progressive	Regressive
Netherlands	Regressive	Moderate
Norway	Regressive	Moderate
Sweden	Regressive	Moderate
Switzerland	Regressive	N/A
United Kingdom	Moderate	Moderate
United States	Moderate	Progressive

Increased grade	No change	Decreased grade	No data

criteria—including a notification of allergens. Furthermore, Austria, Belgium, Denmark, Finland, France, Germany, Italy, the Netherlands, Norway, and Sweden improved from regressive to moderate.

Unlike Australia, Canada, and the United States, 13 of the other countries have significant gaps in their labeling and allergen information regulations. Japan requires that processed (packaged) foods detail the name, list of ingredients, net contents, a best before date, the name and the address of the manufacturer, and storage instructions. Significantly, it does not require nutritional information or a warning about allergens.

Similarly, Austria, Belgium, Denmark, Finland, France, Germany, Ireland, Italy, the Netherlands, Norway, Sweden, and the United Kingdom require only the product name, processing date, nutritional information (but only if a nutritional claim is made on the package), and a list of ingredients. Universal allergen warnings are not required, as EU regulation stipulates only the listing of certain possible allergens in the ingredients list (European Parliament, 2007). However, by acknowledging the risks that certain allergens pose, most of these countries, with the exception of Ireland and the United Kingdom, which were considered moderate performers in 2008, improved their standing.

Australia, Canada, and the United States rank as progressive countries because they require the seven items listed previously (CFIA, 2010).

Ease of Access to Public Health Information

Measurement of the ease of access to public health information was based on a holistic evaluation of the information provided by national-level governments.

Australia, Canada, the United Kingdom, and the United States scored progressive grades because each country's national government provides complete and easily accessible information on their websites. Denmark, Finland, France, Germany, Ireland, Japan, Norway, and Sweden provide some data online; however, the information is generally incomplete or unclear. Austria, Belgium, Italy, the Netherlands, and Switzerland do not provide much, if any, information on their national government websites. As a result, these countries earned grades of regressive.

Comparison with 2008

As illustrated by Table 3.6, Denmark, Finland, France, Germany, Japan, Norway, and Sweden improved their grade to moderate in 2010 compared with 2008. These countries all provided comparably more and more easily accessible information on public health in 2010 than in 2008. Austria, Belgium, the Netherlands, and Switzerland all provided comparatively less information than their peers, and that amount of information tended to fall from 2008 to 2010.

Table 3.6 Ease of access to public health information: 2008 and 2010 grades.

Country	2008 Grade	2010 Grade
Australia	Progressive	Progressive
Austria	Moderate	Regressive
Belgium	Moderate	Regressive
Canada	Progressive	Progressive
Denmark	Regressive	Moderate
Finland	Regressive	Moderate
France	Regressive	Moderate
Germany	Regressive	Moderate
Ireland	Moderate	Moderate
Italy	Regressive	Regressive
Japan	Regressive	Moderate
Netherlands	Moderate	Regressive
Norway	Regressive	Moderate
Sweden	Regressive	Moderate
Switzerland	Regressive	Regressive
United Kingdom	Progressive	Progressive
United States	Progressive	Progressive

Increased grade	No change	Decreased grade	No data

Investigation on Consumer Affairs

The investigation on Consumer Affairs comprised five sections: one outcome and four policies. Their measurements informed the ranking for Consumer Affairs. In addition to the information presented in the analysis of ranking data, this investigation unearthed research that informed the selection of criteria, the importance of these categories, and the background behind these concepts.

Incidences of Reported Illness by Foodborne Pathogens

Reported foodborne-pathogen-caused illness is a concrete, number-based measurement. To ensure that the country-by-country data are collected and measured in the same way, the ideal source of this data is a supranational body such as the United Nations (UN), the World Bank, the World Trade Organization (WTO), or the Organisation for Economic

Co-operation and Development (OECD). Unfortunately, none of these global organizations published data on foodborne pathogens for the 17 countries in this study.

In fact, in January, 2008, the World Health Organization (WHO) launched an initiative to estimate the global burden of foodborne diseases, noting that, at present, there is "no global estimation of the disease burden" (WHO, 2008). The WHO recognized the importance of measuring and monitoring foodborne diseases, but its study highlighted the need for "universally applicable tools" to measure foodborne diseases, while recognizing the limits of comparing different-country data to form a global picture.

As a result, the best 2010 data source is from food safety collection and monitoring bodies in the individual countries. This introduced the possibility of discrepancies in the country-by-country data, since each nation collects data individually and, probably, not identically. Nevertheless, this is the current method for both the EU and the WHO Regional Office for Europe, the only WHO regional office to publish comparative data. Both the European Union and the WHO collect and publish incidences of foodborne pathogens from national health-tracking agencies. With the exception of the data for European countries (drawn primarily from the WHO and supplemented with data collected by the European Union), the comparative foodborne pathogen data were collected from each country's foodborne pathogen monitoring agency.

In addition to being an easily understood and measured output of a country's food safety performance, the rates of foodborne disease incidents is also an important health concern. In 2002, Danielle Nierenberg reported that "food-borne illness is one of the most wide-spread health problems worldwide, and it could be an astounding 300–350 times more frequent than reported" (Nierenberg, 2002). This is a growing problem among first-world countries, including the 17 countries in this study. A frequently cited study notes that there are about 76 million illnesses, 325 000 hospitalizations, and up to 5000 deaths each year in the United States because of foodborne illness (Charlebois and Yost, 2008).

The importance of measuring foodborne illnesses underscores the need to build a global tracking and monitoring system. A 2003 OECD report found that "[*b*]ecause routine surveillance systems vary widely between diseases and between countries, the collected information ... does not allow numerical comparison of data" (Rocourt et al., 2003a, 2003b). As noted, the data for this report were collected from national reporting agencies from the 17 jurisdictions, with the exception of data from European countries, which is derived from the EU and WHO's regional European office. Of these countries, Australia and the United States presented the best and most easily obtainable information.

In Australia, the Australian Department of Health and Aging collects the data and presents it online on the National Notifiable Diseases Surveillance System (NNDSS). Australia reported data for many of the five diseases tracked in this study, including up-to-date information for 2009 and 2010. However, the 2010 data were incomplete and were not used.

In the United States, a simple search of the Centers for Disease Control and Prevention's FoodNet surveillance program reveals complete foodborne illness information for the five pathogens tracked by this study. The United States is also unique in that its information includes reported incidences of *Vibrio* per 100 000 persons. Unlike its neighbor, Canada does not publish easy-to-find or complete information on foodborne illnesses. The data for this study were found through the NESP, which promises to provide "timely analysis and reporting of laboratory confirmed enteric disease cases in Canada" (Canada National Microbiology Laboratory, 2010). The majority of publicly available data used in this study are contained in its 2004, 2005, and 2006 Annual Summaries of Laboratory Surveillance Data for Enteric Pathogens in Canada. The 2006 report is the most current. Canada presumably does collect more up-to-date information; however, the only other retrievable information was found in a 2008 Statistics Canada supplement to a Health Canada report on comparable health indicators, which provided more complete *E. coli* data up to 2006.

In Canada, health jurisdictions are split between federal and provincial monitoring authorities. In fact, this division of responsibilities is a not-uncommon method of organization. In Europe, this system of organization is further supplemented by the European Union, which now requires member states to collect and report on foodborne illnesses annually. In addition to these annual *Trends and Sources of Zoonoses, Zoonotic Agents, and Antimicrobial Resistance in the European Union* reports, the WHO's regional office in Europe (unlike its counterparts responsible for Canada, the United States, Australia, and Japan) collects and reports foodborne illness data through its Computerized Information System for Infectious Diseases (which also covers non-EU members Switzerland and Norway). However, like Canada, many of these countries also delegate the responsibility for collecting foodborne illness data to their subnational governments. For example, Germany delegates the responsibility for monitoring food safety to its states.

With the exception of *Salmonella,* Italy has only recently begun reporting data. This is possibly a result of the EU Zoonoses Monitoring Directive 2003/99/EC, which requires such collection and reporting. Because of the paucity of non-*Salmonella* data for Italy (and because one pathogen does not make a fair comparison), Italy received a grade of not applicable for foodborne illnesses.

Likewise, Japan receives a not applicable grade because there are no data available on the rate of foodborne illnesses in Japan. While research suggests that foodborne diseases are a serious problem, the lack of comparable data removed Japan from the comparison.

Rates of Inspections and Audits

Comparing the rates of inspections and audits among the 17 countries in this study was instructive, in part, because these policies helped to inform the outcome data measured by incidences of foodborne illnesses. Rates of inspections and audits are also key components of the international food safety apparatus. While individual countries or subnational governments may set the policies, the reliability and safety of food is a domestic and international issue, partly due to the increase in international trade.

The OECD attributes some of the increases in global foodborne diseases to the rise of globalized food trade. In addition to more rapidly spreading diseases from infection-prone jurisdictions, the long-distance travel of goods increases the risks that food may become infected or that new infections can spread to an otherwise uninfected population (Rocourt et al., 2003a, 2003b). Part of the risk for foodborne diseases may be a result of different rules, regulations, and oversight mechanisms in different jurisdictions. According to the UN's Food and Agriculture Organization (FAO) and the WHO, "the harmonization of food standards is generally viewed as contributing to the protection of consumer health and to the fullest possible facilitation of international trade" (Secretariat of the Joint FAO/WHO Food Standards Programme, 2006a).

To address these concerns, the global CODEX Alimentarius Commission (CODEX) outlines international regulations and guidelines to assure the safe production and transportation of food (Secretariat of the Joint FAO/WHO Food Standards Programme, 2006b). In 1961, the FAO and WHO collaborated on the program's creation in an attempt to provide leadership in the area of international food safety, to provide rules for safe food, and to standardize those rules to reduce international trade barriers. For countries included in this study—high-income, first-world industrial nations—CODEX rules complement existing national policies. However, these types of countries report that they use CODEX standards only as a measurement of the safety of food imports.

Food Safety Education Programs

With the exception of Norway and Switzerland, six of the countries that received a grade of regressive are members of the European Union. Furthermore, both Norway and Switzerland have agreements with the European Union regarding food and food safety. The European Union, through the European Food Safety Authority (EFSA), has a supranational farm-to-fork plan and food safety organization. Yet the involvement of the EFSA may contribute to subpar national education programs, as its mandate includes the research and communication of food safety information to the European public. While this hypothesis does not explain the disparity between the United Kingdom (progressive), France (moderate), and Germany (regressive)—the three largest economies in the European Union—the shift toward supranational food safety programs in the European Union may help account for the relative absence of food safety education programs at the national level in smaller EU member states.

Labeling and Indications of Allergens

Like food safety education programs, regulations governing the labeling and indications of allergens have become the domain of the European Union as opposed to its member nations. Even outside of the European Union, Norway (a non-EU member) has adopted EU regulations (Norway Lovdata, 2010). As a result of this supranational harmonization, the homogeneity of food labeling regulations in Europe, coupled with the CODEX standard for labeling of prepackaged foods (Codex Alimentarius Commission, 2008), may render future evaluations in this category more difficult because of decreasing differences.

Discussion

In the Consumer Affairs ranking, Canada's performance is respectable. Nonetheless, some improvements ought to be considered. Methods by which governments and industry alike engage with consumers are becoming increasingly important. Systemic events have triggered waves of fear among the general public, and trust between government and the public is frail. For example, the swine flu epidemic recalibrated that relational trust. In 2009, Egypt began slaughtering its roughly 300 000 pigs as a precautionary measure against the spread of swine flu, even though no cases have been reported in either humans or animals in that country (Kerr, 2009). Underlying religious reasons led to this decision, but the policy is still considered by many to be extreme. At least 12 countries, from China to Russia to Ukraine, set bans on all pork product imports (Garrett, 2009). Even Canada saw its share of panicky assessments. Pork inspections increased in an attempt to ease any concerns consumers had in the aftermath of the global influenza A H1N1 outbreak. Even so, there was no evidence that swine flu can be transmitted through eating meat from infected animals. Worse, no scientific data suggest the strain in its entirety has any link with hogs. In the meantime, hog futures were down at the time, and the hog industry was losing millions (Kerr, 2009). From an epidemiological standpoint, the menace was real. But when it comes to food safety, the "swine flu" story was really nothing more than a sign of our fearful times.

All around the world, panic occasioned by outbreaks of the influenza A H1N1 virus is incommensurate with the risks of contracting it by eating pork. Public officials are making unfounded decisions based on trivial information at best. Many elements have triggered this phenomenon. The obvious culprit, for lack of a better term, is the media, but there is a lot more going than mere fear-driven reporting.

Many criticized the media for the insufficient information and/or misinformation with which they bombard the general population in times of crisis. Nevertheless, misapprehension of risk by media and news audiences is not exclusively responsible for disproportionate fear or panic. Still, panic is more likely to occur when a threat receives substantial media coverage. In a crisis, media need instant access to data that may not yet exist, specialists and authorities who are not available, or assertions about issues that appropriate sources may not feel prepared to make. Like consumers and trade officials, the media, too, have to make crucial decisions with weak evidence. In addition, because public understanding of science is often limited with respect to such crises, it is challenging for the media to sufficiently clarify complex phenomena so that the public can understand them.

However, the media can hardly be blamed, given the multifaceted state of affairs during food crisis events. In essence, the media coverage for the influenza A H1N1 is both frightening and functional. It cultivates fear and apprehension but provides an opportunity for the public to become educated about the issue. The media is fuelling the problem, perhaps, but it has reason. Currently in Canada, no public authority can play a more effective role than the media can in times of a public health crisis. The media can operate virtually in real time in this day and age, whereas public authorities must react to incidents painstakingly. Unlike the media, from whom the public is willing to accept a certain degree of error and backtracking, public health authorities simply cannot prematurely release information that could have so devastating an effect as to, for example, prompt the unnecessary slaughter of hundreds of thousands of perfectly safe and healthy livestock.

What was interesting during the H1N1 crisis, at least in the Canadian media, was the frequent repetition that pork is safe, that there is no evidence this particular flu is spread by pigs or by pork products, and so on. Nevertheless, pork futures dropped and healthy pigs slaughtered in unaffected parts of the world as a "precautionary measure." This may have less to do with real panic over the epidemic and more to do with the way that food trades work.

For food trades, among other things, borders are usually only selectively opened. Borders serve important roles in modern economies. They safeguard sovereignty of nations, impose economic protection, and facilitate the screening out of undesirable products. Regrettably, embargoes and restrictions occur almost on a daily basis for all sorts of reasons, justified or not. The influenza A H1N1 is the latest excuse some countries are using. It is just part and parcel of the normal trading process.

Nonetheless, these measures foster market fear, which leads to unwise behaviours. Pork is the cheapest protein available, and many developing countries depend on its accessibility and affordability to feed their populations. Decisions to restrict pork imports will actually have a devastating impact on food retailers, the transportation sector, restaurants and, most importantly, consumers. Economic multiplier effects, coupled with public panic concerning the infectivity of influenza A H1N1 in people, may also create a pervasive atmosphere of fear and pessimism that could aggravate the current economic downturn felt in the global economy.

What really triggered this public health panic was the fact that the public is now more prone to accept risk prevention measures than ever before, especially in matters involving food. BSE, foot and mouth, major *Salmonella*, *Listeria*, and *E. coli* outbreaks have forever altered risk perceptions of consumers worldwide. Even the succession of less serious food recalls and food-related nuisances have taken their toll on the world population, and public authorities know this. Nonetheless, many argue their decisions may make matters worse, not better.

It is difficult to blame anyone, therefore, for the hog industry's woes. The media, the public, and the food trades are all inclined to respond aggressively and perhaps prematurely to such events. The baseless correlation between the current pandemic and the hog industry simply exists in people's minds, and it may be too late to avert major losses. Chicken farmers in some parts of the world had to face the same problem a few years back with the avian flu spread. Media do have a role to play in food crises, but global events related to food are inherently fluid and the message should constantly change. The disconnect between hogs and influenza A H1N1 should become evident in a relatively short time, as the epidemic itself seems to be mellowing. Whether such revelation will occur in time to prevent serious damage to agriculture sectors worldwide remains to be seen.

Other challenges are becoming more palpable for the food safety systems from around the world, and Canada is certainly not immune. In the lag between awareness of a food safety problem and identification of its source, history has repeated itself, with harmful consequences for

Canadian consumers and their food industry. Soon after Toronto-based Siena Foods recalled five meat products, health officials confirmed that none of the reported five deaths were connected with the recall (Smith, 2010). But further economic damage already occurred via a class action suit against the manufacturer of the tainted products. We witnessed another case of delayed identification during the mad cow crisis of 2003, when it took inspectors 4 months to trace the diseased animal to its original herd. More recently, in 2008, Ontario Health officials diagnosed people infected with listeriosis 2 months before the outbreak was linked to the Maple Leaf plant in the Toronto area. The deaths of 22 Canadians in these crises are tragic enough; yet we must also contend with widespread public anxiety in the wake of these outbreaks. We are clearly unable to adequately track and monitor risks in the food industry.

What we have now are post hoc arrangements among competing stakeholders, consumers who often lack basic information, and regulatory bodies, particularly the CFIA, which must somehow instil collaboration and trust among all parties. The results, in the event of a food crisis, have been disorganization, inefficiency, and uncertainty.

Our current practices need to become more organized, but there are contrasting views about how to best shape our food safety systems. Some groups, often led by public spokespersons, are calling for more food safety inspectors. The CFIA, on the other hand, claims its inspectors are sufficiently deployed and that the frequency of inspections is adequate. In fact, the CFIA already employs well over 5000 employees and 200 inspectors. More inspectors on the field will not necessarily translate into more direct testing of food products. More likely, it would mean that companies must contend with more bureaucracy. It simply does not follow that hiring more inspectors will make our food safer.

Instead, our objective should be a comprehensive food traceability system, a task that demands high organizational flexibility from the entire food industry. The endeavor is more than worthwhile, even though implementation will be challenging. Unfortunately, a competitive environment in which consensus building is difficult and even basic infrastructural agreements among stakeholders may be hampering regulators. Additionally, keeping in mind past failures with forced cooperation in the food industry, it is necessary to promote a willingness to cooperate and to foster a supply chain-oriented paradigm.

Food traceability systems can achieve this integration, and the CFIA is the perfect forum in which to build such highly needed cooperation. The CFIA, despite previous difficulties, must redouble its efforts to rally the food industry before and after food safety crises. Risk management across the food industry can be successful only if all major stakeholders and production chains are called to account.

But building better food safety systems requires more than industry collaboration. It should also address the lack of transparency displayed by both regulators and industry. Many Canadians believe we need more inspectors because of misunderstandings about what an inspector does and how the industry operates. In all of our recent food crises, harm to consumer perception was aggravated by lack of transparency in food industry organizations. Since consumer perception of food hygiene and safety are some of the key drivers of food safety policy, it is vital that consumers understand more fully how the food industry operates and is regulated.

The CFIA needs to become a better risk communicator. In the absence of a mandatory disclosure system, the public is currently unable to obtain information about food safety records and scorecards generated by inspections. This has to change as it raises questions about the robustness of the regulatory process. Consumers deserve to be better educated about food safety matters, and once educated they will repay the food industry with trust.

The debate should not be about the number of inspectors hired by the CFIA, because it is much too simplistic for the food industry's segmented realities. Rather, the debate should be about how to build food safety systems that serve the health requirements and information needs of Canadian consumers. For years, we have been chasing the wrong targets because we have not been able or willing to spur cooperation and accountability in a competitive food industry. In its responses to food safety crises, the food industry would benefit from statutory regulations that require traceability from farm to plate. Further, implementing nonregulatory quality assurance programs would bolster consumer confidence. These approaches emphasize traceability, transparency, and accountability within food production and have previously acted as catalysts for more collaborative behavior. More collaboration within the food industry, especially between regulators and industry, could make an extensive difference to Canadians' physical and mental well-being.

4

Biosecurity

Today's dangers are often less visible than one can expect. As food distribution systems become more complex, it is becoming challenging to pinpoint potential sources of contamination. The events of September 11, 2001, certainly changed how we manage risks, or perhaps how we manage public fears. Most importantly though, it has changed how we perceived risks. Since then, many industrialized countries have launched initiatives that are designed not only to help companies comply with requirements of bioterrorism threats but also to move beyond that into expanded recordkeeping that certain groups and governments hope will help improve the safety of their domestic food supply or at least speed up the recall efforts in the event tainted products reach the market.

It is difficult not to recognize the emergence of a new approach in how nations frame risk. In particular, the threat to the food production, processing, and distribution system can no longer be limited to accidental or perhaps unintentional. Even though most, if not all, food safety incidences are human induced, some may be ill intended and intentional. Some have argued certain groups may intentionally harm a massive number of consumers (Halweil, 2004). Agricultural systems have been considered as prime targets for any terrorist groups wanting to reach a great number of citizens. As a result, governmental regulators also are becoming aware of the vulnerability of the food supply chain as a target for bioterrorism. How that risk will manifest itself remains unclear, but how one defines the scope of a given system to properly identify potential risk is becoming key (Mooney and Hunt, 2009).

The will to regulate by many governments is higher than ever and extends beyond agriculture. The trend toward the concentration of certain agricultural businesses into the hands of a small number of large conglomerates gives rise to yet another need for government oversight. Many feel the regulation of food safety must address the systemic risk posed by the global consolidation and vertical integration of agrifood

Food Safety, Risk Intelligence and Benchmarking, First Edition. Sylvain Charlebois.

businesses and imposing agrifood networks. We have seen a particular increase in the creation of megacorporations in the Western world.

The use of pesticides and chemicals is also a matter of great importance. Biosecurity protocols aim to reduce or eliminate animal pests and plants that may harbor disease. In doing so, it may influence the levels of chemicals consumers are exposed to. Pesticides and herbicides need to be properly used. They are diluted in water for use when they are in the concentrated form, and care must be taken when handling as they are potent poisons when in the concentrated form. This may enhance the risks for consumers. Much research has been conducted to better understand the repercussions on agricultural chemical use over the long term. We know that the use of agricultural chemicals can impact the level of safety that some products may provide to consumers in general. As a result, many countries consider reduced exposure to agricultural chemicals that are a direct contributor to improved food quality. Agricultural chemicals can impact human health in many ways. Children have been known to be exposed to chemicals, and some studies have identified risk levels that do concern some. Underage workers in developing countries are also a source of concern when considering the effects of agricultural chemicals. It has been recognized that children of agricultural workers are at even greater risk of the serious health effects of chronic pesticide exposure (Lucas and Allen, 2009). Even in developed countries, children living in rural regions or who are exposed to chemicals may be at risk as well. Proximity to fields and insentient overexposition are genuine threats. Some have also argued that there is a longstanding concern in occupational hygiene that exposures to neurologically active chemical compounds may influence mood and behavioral impairments that increase the risk of suicide, and agricultural chemicals would fit in this category (Mustard et al., 2010). These are highly lauded benefits by many stakeholders concerned with food safety. Therefore, the usage of agricultural chemicals by countries is an important food safety variable.

Beyond pesticide use and residue, the use of other chemicals has become a concern. Without standards, the agrifood industry may expose consumers to unwarranted risks. In developing countries like China, they are struggling to control the use of agricultural chemicals since the agrifood business's focus is mainly production. Systemic risk considerations and the evaluation of chemical residues are not as much of a priority compared to developed countries in which consumers have different expectations. China's practices with agricultural chemicals have attracted the world's attention since 2007, with specific cases related to melamine in milk products. Since then, with the assistance of the World Trade Organization (WTO), Chinese-based farms which use pesticides, veterinary medicines, feed additives, and fertilizers that have been banned or limits specified for their application have become priority areas for improvement. Standards are slowly being harmonized around the world. Eliminating the use of banned pesticides, other agri-related chemicals, and feed additives and strengthening the inspection of imports and exports of produce and other food products are also areas of significance.

Generally speaking, food contaminants can be generally characterized as pathogens, biologically produced toxins in plants or produced by pests and toxins from other sources such as agrochemical residues and environmental pollutants. Of course, the most popular means of addressing uncleanness with pathogens is through improvements in sanitation and hygiene. What we have seen in the last quarter of a century was an increase in pesticide use in developing countries as they transition out of labor-intensive agricultural practices (Hoffmann, 2009). Loose regulation, lax enforcement, and price sensitivity all often lead to the consideration of older, less expensive, nonpatented, and generally more acutely toxic and environmentally persistent agents (Ecobichon, 2001). Through the years, contamination with agricultural chemicals is likely to be a much significant problem in developing countries than in Western-based economies. This essentially is due to harsher regulations and strict enforcement methods. But, ironically, some economies perceive the use of agricultural chemicals as an occupational hazard rather than a food safety issue. A case can be made that the use of agricultural chemicals can certainly become a food safety concern. Modern economies are compelled to measure and monitor the use of chemicals for agricultural purposes. Consumers now do expect strict monitoring from food safety agencies. In developing countries, the situation is very different, and the health of consumers may sometimes be compromised. The lack of emphasis on managing contaminants may harm consumers, but negative externalities are not recognized until many years later. In addition, farmers also tend to be less proactive since access to information is limited.

Developed countries have seen their share of exposure to contaminants. As many farmers in developed countries are generally highly dependent on chemical fertilizers and pesticides in growing commodities, chemicals remain vital for them, and these practices have been adopted from generation to generation. For example, agricultural pesticide contamination of sediments from five Mississippi Delta oxbow lakes and their effects and bioavailability to *Hyalella azteca* were assessed during the fall season. The fall is recognized as a low-application season. Some pesticides were detected (Lizotte Jr. et al., 2009). More endowed countries do have the capacity and technologies available to properly monitor agricultural chemicals. One such technology is radio frequency identification (RFID) tags. The vital information that can be stored on a RFID tag is interesting. These tags can store information such as country of registration, chemical type, unique registration number of agricultural chemicals, container size, specific gravity, unit of measure, and a digital signature (Peets et al., 2009). Some experiments have been conducted and results were conclusive. The use of such technology allows a supply chain to better manage residues generated from agrichemicals. It is important thought to consider that agricultural chemicals are of economic importance as well for farmers. Given that farming is considered as one of the most capital-intensive industries, the cost of inputs is of significant importance. Prices of agricultural chemicals fluctuate from time to time.

The use of agricultural chemicals is particularly crucial for countries acting as a competitive hub and prominent exporters of agricultural commodities. In the future, many of these countries may encounter a greater demand for clean and safe-to-consume agricultural products, particularly from its key buyers, comprising Japan, South Korea, and Europe. These buyers are often dependent on food imports for a variety of products. Consumers have also changed their eating habits due to growing evidence that some types of food may represent higher perceived health risks more than others. More and more consumers are becoming vegetarians, but they remain exposed to agricultural chemicals. Kathpal and Kumari (2009) demonstrated that vegetarians consumed an unreasonable level of pesticides. They found that actual daily intake of lindane in two samples and endosulfan in four exceeded the acceptable daily intake. In response, some farmers have chosen to grow organic products because of health and safety concerns (Cranfield et al., 2010). Environmental issues are the predominant motives for conversion, while economic motives are of lesser importance.

Bioterrorism

Bioterrorism and food crimes in global food systems exist. The deliberate contamination of food is an emergent problem in many countries. For example, in China, a food distributor poisoned its competitor's food products in pursuit of increasing profits and local market positioning. This incident created several victims from across the country (Johnson, 2008). It was considered as a terrible act of food sabotage (Anonymous, 2008).

The investigation into biosecurity comprised one outcome and one policy: the rate of use of agricultural chemicals and whether a country has a bioterrorism strategy. Countries were ranked based on their performance in these criteria on a sliding scale from superior to poor.

Within the three grades (superior, average, and poor), countries were ranked against their peers through a comparative study of pesticide need and bioterrorism risk.

In grading and ranking the countries, consideration was given to particular country-specific variables. For example, Japan (which uses comparatively more pesticide than Canada) has a greater need for pesticides because of its climate. Similarly, Finland (which began the process of creating a counterterrorism strategy in 2010) is unlikely to have the same risk factors as the United States, which lowers the necessary State–Pressure–Response system to perform, comparatively, well.

As illustrated in Table 4.1, Denmark, Finland, Norway, and Sweden earned biosecurity grades of superior. Each of these countries has extremely low averages of pesticide use compared (in context) with the

Table 4.1 Ranking for Biosecurity.

			2008 Comparison	
Rank	**Country**	**Grade**	**Rank**	**Grade**
1	Denmark	Superior	1	Superior
1	Finland	Superior	5	Superior
3	Norway	Superior	8	Superior
4	Sweden	Superior	9	Superior
5	Austria	Average	4	Superior
5	Germany	Average	6	Superior
7	Switzerland	Average	3	Superior
8	Japan	Average	2	Superior
9	United Kingdom	Average	10	Superior
9	United States	Average	13	Average
11	Australia	Average	11	Average
11	Canada	Average	14	Average
13	Ireland	Poor	7	Superior
14	France	Poor	12	Average
15	Belgium	Poor	17	Poor
15	Italy	Poor	15	Average
17	Netherlands	Poor	16	Poor

Increased grade	No change	Decreased grade	No data

other 16 countries. Australia, Austria, Canada, Germany, Japan, Switzerland, the United Kingdom (UK), and the United States scored grades of average for their medium but not too high use of chemical products and generally reasonable bioterrorism programs. Belgium, France, Ireland, Italy, and the Netherlands finished with rank of poor for their unusually high use of pesticide and insufficient bioterrorism strategies.

Comparison with 2008

Comparisons between 2008 and 2010 are difficult because of the addition of the bioterrorism strategy criteria. Nevertheless, when considering this new metric and the data on pesticide use, Austria, France, Germany, Ireland, Italy, Japan, Switzerland, and the United Kingdom all fell in grade. With the exception of Ireland, no country fell by more than one level. Generally, their performance in this category was underwhelming. In particular, Austria, Germany, and Italy all lost marks for their inexplicable lack of bioterrorism plans.

Analysis of Ranking Data

Rate of Use of Agricultural Chemicals

This biosecurity metric measured the amount of pesticides used by each country. This comparison illustrated the country-by-country rates of use by the number of tonnes of pesticide used on each hectare of agricultural land using a 10-year (1995–2004) average.

The average pesticide use per hectare of agricultural land was calculated for each country by dividing the published rate of pesticide use (in kilograms of active ingredients) by the amount of agricultural land (arable and permanent crop area). These year-by-year rates were then averaged over the 10-year period.

Australia, Austria, Canada, Denmark, Finland, Germany, Norway, Sweden, Switzerland, and the United States all received grades of progressive, each using less than an average of 4 kg of pesticide per hectare of agricultural land. Belgium, France, Ireland, Italy, and the United Kingdom received grades of moderate for using between 4 and 8 kg of pesticide per hectare. Japan and the Netherlands received grades of regressive for using over 8 kg of pesticide.

Comparison with 2008

As illustrated in Table 4.2, Australia, Belgium, Canada, and the United States all improved their 2010 grade compared to 2008. Compared with their peers, each of these countries used less pesticide per hectare of

Table 4.2 Rate of use of agricultural chemicals.

Country	2008 grade	2010 grade
Australia	Regressive	Progressive
Austria	Progressive	Progressive
Belgium	Regressive	Moderate
Canada	Regressive	Progressive
Denmark	Progressive	Progressive
Finland	Progressive	Progressive
France	Moderate	Moderate
Germany	Progressive	Progressive
Ireland	Progressive	Moderate
Italy	Moderate	Moderate
Japan	Progressive	Regressive
Netherlands	Regressive	Regressive
Norway	Progressive	Progressive
Sweden	Progressive	Progressive
Switzerland	Progressive	Progressive
United Kingdom	Moderate	Moderate
United States	Regressive	Progressive

Increased grade	No change	Decreased grade	No data

agricultural land than they did in 2008. This is a noteworthy change compared with 2008, as all of these countries were graded regressive in 2008. In 2010, Australia, Canada, and the United States each jumped two grades by reducing their pesticide use per hectare of agricultural land to below 2 kg. Belgium raised one grade by reducing its use below 7 kg/ha.

Ireland and Japan scored lower in 2010 than their 2008 grades. Ireland, which scored a progressive grade (less than 4 kg of pesticide per hectare of agricultural land) in 2010, fell to moderate using a 10-year average of 5.25 kg. Japan dropped two grades, falling from progressive in 2008 to regressive in 2010 with an average of 17.75 kg of pesticide use per hectare of agricultural land. This is a particularly severe drop because its 2010 average use is well ahead of the 17-country average (4.39 kg/ha) and almost 7 kg above the Netherlands, the 16th-place finisher.

While Canada scored a progressive grade, its average and that of Australia were calculated based on far fewer data points than the other countries: Canada has a 2-year average (1999 and 2000), while Australia has a single year (2000) average.

Bioterrorism Strategy

This study generally considered food safety outcomes and the policies that create them, which were based on the accidental, natural or otherwise-unintended threats. However, the risk of bioterrorism—malicious, intentional attacks against a country's food system—has increased. The 2007 World Health Report lists "the natural accidental and deliberate contamination of food" as "one of the major global public health threats in the 21st century" (WHO, 2008). According to the UK's Centre for the Protection of National Infrastructure (CPNI):

> The public and businesses within the food and drink sector now face a different threat—that of malicious attack, especially by ideologically motivated individuals and groups. This threat will manifest in a way which reflects the motivation and capability of these people. It will not follow the statistically random, and therefore predictable patterns of familiar "hazards" (UK Centre for environmental protection of National Infrastructure, 2010).

As a result, bioterrorism food safety policies must be adaptable, comprehensive, and effective. At the very least, countries must include bioterrorism plans as part of their food safety apparatus in order to ensure the safety of the food supply from accidental and intentional threats. According to the WHO, while measure should be "proportional to the size and nature of the threat" (in other words, countries seen to be at a greater risk to bioterrorism should have more established policies), the basic requirements are the same for all countries: cooperation between government agencies and the involved industry, as well as existing bioterrorism systems and plans (WHO, 2008).

As a new addition to the 2010 study, each of the 17 countries included was evaluated based on its bioterrorism strategies. If a country included bioterrorism in its food safety or counterterrorism plans, it received a grade of moderate. Those countries with comprehensive plans with established intergovernmental cooperation (including the involvement of law enforcement and security agencies) received grades of progressive. Countries without clear bioterrorism strategies received grades of regressive.

In 2010, the United Kingdom and the United States received grades of progressive for including bioterrorism within their counterterrorism and food safety organizations. Australia, Canada, France, Ireland, and Japan received grades of moderate for recognizing the importance of preventatives strategies and response plans, despite the absence of clear coordination and interagency cooperation. Austria, Belgium, Denmark, Finland, Germany, Italy, the Netherlands, Norway, Sweden, and Switzerland all provided unclear, incomplete, or absent bioterrorism strategies.

Comparison with 2008

As illustrated in Table 4.3, there are no comparable 2008 grades, as the 2008 food safety world ranking study did not consider bioterrorism programs. Nevertheless, interesting interregional patterns emerged.

Generally, the non-European countries performed adequately, with Australia, Canada, and Japan scoring at the moderate level. The United States, which reorganized its counterterrorism agencies and programs after the September 11, 2001, terrorist attacks, had the best and most

Table 4.3 Bioterrorism strategy.

Country	2008 Grade	2010 Grade
Australia	N/A	Moderate
Austria	N/A	Poor
Belgium	N/A	Poor
Canada	N/A	Moderate
Denmark	N/A	Poor
Finland	N/A	Poor
France	N/A	Moderate
Germany	N/A	Poor
Ireland	N/A	Moderate
Italy	N/A	Poor
Japan	N/A	Moderate
Netherlands	N/A	Poor
Norway	N/A	Poor
Sweden	N/A	Poor
Switzerland	N/A	Poor
United Kingdom	N/A	Progressive
United States	N/A	Progressive

integrated bioterrorism strategy—earning that country a grade of progressive. The United Kingdom also had a well-developed bioterrorism program, including that risk with its national plan and integrating its infrastructure protection efforts with its preventative law and security agencies. While its closest neighbors (France and Ireland) had adequate bioterrorism strategies, the remainder of Europe fared poorly.

There was little data available on Austria, Belgium, Denmark, Germany, Italy, the Netherlands, Norway, Sweden, and Switzerland (earning these countries grades of poor). Finland has only begun creating a general counterterrorism program (which may or may not include bioterrorism). Moreover, while the European Union (EU) acknowledges the risk of bioterrorism (beginning in 2002 and 2003 it participated in the US efforts to increase security against bioterrorism attacks), generally, it attempts to increase cooperation between its member states and to facilitate coordinated bioterrorism responses. This leaves the primary and principle responsibility for bioterrorism to the member states. With the exception of the United Kingdom (which has an excellent strategy) and perhaps France and Ireland (with their adequate programs), Europe as a whole requires a significant increase in its bioterrorism prevention and management strategies.

Investigation on Biosecurity

The investigation on biosecurity comprised one outcome (the rate of use of agricultural chemicals) and one policy (a country's bioterrorism strategy). In addition to the absolute 17-country ranking of pesticide use, this research unearthed information that may help to inform the absolute rate of pesticide use and the change (between 2008 and 2010) in some of the countries' scores. Moreover, as an issue of increasing importance in the early twenty-first century, bioterrorism plans and strategies are an important link in a country's food safety system. While all countries are not equally susceptible or equally as likely to become targets or victims of terrorism, good preparation is an important metric to consider.

Rate of Use of Agricultural Chemicals

The rates of pesticide use is a concrete, number-based measurement. To ensure that the country-by-country data were collected and measured in the same way, the ideal source of this data is a supranational body, like the Organisation for Economic Co-operation and Development (OECD). The data from this section were taken from a 2008 OECD report, *Environmental Performance of Agriculture in OECD Countries Since 1990.*

From this report, this study took two variables: the amount of agricultural land (measured by the number of hectares of arable and permanent crop areas[1]) and the amount of pesticide used (measured by the use and sale of the chemical agents). Table 4.4 provides detailed information on the per-agricultural hectare use of pesticide by each country.

Overall, the countries considered in this study are among the major world food suppliers (OECD, 2008). In part, this may be a result of their, generally, large geographical area and established agricultural industries that produce food for domestic and international consumption. It is important to recognize that some countries are using less land for agriculture. However, changes in gross agricultural land do not, necessarily, mean that these countries are producing more or less food. In fact, the OECD study found that Australia, Canada, and the United States are increasing production while maintaining relatively stable agricultural areas, while Norway is decreasing its production but increasing its land use. On the other hand, Japan, the Netherlands, and the United Kingdom are shrinking their production and agricultural area. According to the OECD, overall growth in agricultural production is likely to be a result of increased yield. As a result, pesticide use ("chemical inputs") per hectare may increase along with this new yield.

Measuring the use of pesticide is difficult as few countries keep data on use, and there are few comparative metrics for the collection of data. As a result, the OECD uses the sale of pesticides as "a proxy" for use. Because of the difficulty of standardizing incomplete national data, this study relied on the OECD report to ensure that the data are treated and measured similarly. While the OECD found that overall pesticide use fell (a conclusion that considers countries outside the scope of this study), that study also found variances in the country-by-country usage data. This can be, in part, attributed to declining crop production (Denmark and Switzerland), more aggressive reduction targets (Denmark and the Netherlands), increased taxes on pesticide use (Denmark and Norway), better targeting use (Austria, Norway, and Switzerland), or organic farming (Austria, Denmark, Finland, Japan, and Switzerland). However, in addition to the reductions of some countries, others continue to dominate the overall use of these products. Nevertheless, the gross measurement of pesticides used does not necessarily quantify the possible environmental or health risks because the OECD trend is toward "less environmentally harmful" products.

---

1 Arable land is defined as land "cultivated for crops like wheat, maize and rice that are replaced after each harvest." Permanent crop areas are land "cultivated for crops like citrus, coffee and rubber that are not replanted after each harvest."

Table 4.4 Kilograms of pesticide used per hectare of agricultural land.[a]

Country	1995	1996	1997	1998	1999	2000	2001	2002	2003	2004	Average over 10 Years
Australia	—	—	—	—	0.75	—	—	—	—	—	0.75
Austria	2.28	2.39	2.49	2.27	2.32	2.44	2.15	2.11	2.32	2.27	2.31
Belgium	7.52	7.48	7.26	7.39	6.91	7.03	6.23	6.49	6.32	6.60	6.92
Canada	—	—	—	—	1.05	0.92	—	—	—	—	0.99
Denmark	2.07	1.57	1.55	1.52	1.32	1.29	1.35	1.38	1.27	1.26	1.46
Finland	0.49	0.44	0.48	0.54	0.52	0.53	0.65	0.73	0.75	0.67	0.58
France	4.31	5.03	5.64	5.58	6.17	4.84	5.09	4.21	3.81	—	4.96
Germany	2.53	2.66	2.55	2.78	2.51	2.53	2.32	2.46	2.45	—	2.53
Ireland	5.64	4.28	5.61	6.03	5.15	5.23	5.31	5.09	4.94	—	5.25
Italy	6.70	6.79	7.55	7.57	7.25	7.13	6.92	8.66	7.92	—	7.39
Japan	19.72	19.30	19.68	18.99	18.55	18.43	15.76	15.91	15.65	15.50	17.75
Netherlands	12.91	11.29	12.26	12.54	11.98	11.32	9.37	9.25	8.79	9.92	10.96
Norway	1.04	0.79	0.84	1.06	0.90	0.43	0.59	0.93	0.79	0.99	0.83
Sweden	0.44	0.54	0.57	0.58	0.62	0.61	0.64	0.64	0.78	0.37	0.58
Switzerland	4.45	3.85	3.66	3.54	3.43	3.56	3.52	3.44	3.35	3.15	3.60
United Kingdom	5.63	5.82	5.64	5.73	5.90	5.66	6.34	7.25	5.54	—	5.95
United States	1.88	2.01	1.89	1.83	1.79	1.83	1.74	1.74	1.74	—	1.83
									17-country average: **4.39**		

a) Table generated from the *Environmental Performance of Agriculture in OECD Countries Since 1990* report. The kilograms of pesticide used per hectare of agricultural land is calculated by taking the report's country-by-country arable and permanent crop area (in hectares) and dividing it by the total kilograms of pesticide use (tonnes of active ingredients divided by 1000).

As a result, while the measurement of pesticide use per hectare is a useful indicator of biosecurity, this measurement does not necessarily reflect the whole impact. For example, a country that increases its use of land (in hectares) to grow the same amount of food without increasing its gross use of pesticide would have a lower average after this change than a country that has increased its yield and pesticide use on the same amount of land. Moreover, these pesticide statistics do not reflect the specific risks associated with different chemical agents.

This finding is also highlighted in the E.U.'s 2007 environmental assessment, which found that "less toxic ingredients" in pesticides were "likely to have reduced the general environmental pressures ... although their use remains high in certain farming systems." This included Western Europe and countries with higher levels of irrigation farming (which require more pesticide) (European Union European Environmental Agency, 2007). This 2007 report used United Nations Food and Agriculture Organization (FAO) data from 1990 to 2001, which may echo the paucity of recent data found for this study as well. While the EU report does not provide specific figures, the EU-15 and EFTA-4 (a 19-country list that includes 11 EU members that is in this study and European Free Trade Association (EFTA) members Norway and Switzerland (European Free Trade Area, 2010)) showed a 19-country average of between 2 and 3 kg of pesticide use per hectare of agricultural land between 1990 and 2001 (European Union European Environmental Agency, 2007). This compares with an average of about 4 kg of pesticide per hectare in European countries included in this study based on the OECD data. This difference makes sense considering that this study comprises only Western European countries with the population and landmass to maintain significant agricultural industries.

Discussion

The general rule for evaluating each country's bioterrorism strategy was to understand the role that preparations to prevent or manage bioterrorism were incorporated with general food safety and counterterrorism organizations and strategies. The key metrics were whether a country included bioterrorism as a distinct element of its counterterrorism plan, an evaluation of the level of comprehensiveness that this plan had in the overall food safety apparatus, and whether this strategy included the integration of all necessary organizations, including law enforcement and/or security.

The United Kingdom's counterterrorism strategy includes the safety of the food supply under critical national infrastructure. These measures can also be interpreted as being economically motivated. It should be noted that bioterrorism is not solely about mitigating potential threats but it is also about reducing risks of intentional regular business disruptions. Included in this program are the identification of vulnerabilities, on-site security enhancements, and dialogues with industry (UK Secretary of State for the Home Department, 2010). The United Kingdom's CPNI specifically combines government and industry expertise to attempt to harden national infrastructure (including the food chain) from bioterrorism. This activity ("protect") falls under the United Kingdom's four-step counterterrorism strategy: pursue, prevent, protect, and prepare. As the "pursue" component includes police and security agencies, the United Kingdom's bioterrorism strategy was considered to include both a plan as well as the close involvement of law enforcement and intelligence agencies.

The United States also has a detailed and comprehensive partnership between its food safety organizations (the United States Department of Agriculture (USDA) and the Food and Drug Administration (FDA)) and law enforcement and security agencies (the Federal Bureau of Investigation (FBI) and the Department of Homeland Security (DoHS)). Shortly after the September 11, 2001, terrorist attacks on the United States, the American government passed the *Public Health Security and Bioterrorism Preparedness and Response Act*, which increased the regulatory power and responsibilities of food safety organizations. As part of this bioterrorism mandate, the USDA, FDA, DoHS, and the FBI created the Strategic Partnership Program Agroterrorism Initiative (SPPA) initiative in 2005 (USDA, 2010) to "provide government and industry with a more complete sector-wide perspective of food and agriculture defense in the sector" (USDA, 2007). In collaboration with subnational and industry partners, the SPPA studies food industry vulnerabilities, identifies possibilities for attacks, and builds relationships between "Federal, State, and local law enforcement and the food and agriculture industry along with the critical food/agriculture sites."

Like the United Kingdom, Australia's 2010 Counter-Terrorism White Paper (*Securing Australia—Protecting Our Community*) includes the country's Critical Infrastructure Protection Program (Australian Government, 2010), which includes the Food Chain Infrastructure Assurance Advisory Group. This organization forms one of the nine infrastructure groups that are included in a government-to-government-to-industry network of information and policy. Participation in this group gives subnational and industry participants the opportunity to communicate and offer recommendations to the national government. In 2008, the *Independent Review of Australia's Quarantine and*

Biosecurity Arrangements Report to the Australian Government highlighted the risk of agriterrorism (Beale et al., 2008) and recommended that Australia's biosecurity functions be combined in a new national biosecurity authority. (In addition to information gathering and enforcement responsibilities, this organization would also liaise with police and security agencies.) As of mid-2009, the Australian government has committed to enacting all 84 of the report recommendations and has reorganized its biosecurity apparatus (Australia Department of Agriculture, Fisheries and Forestry, 2009). Australia is currently moving toward a comprehensive bioterrorism plan but has not fully developed a stand-alone organization.

Canada does not have a comprehensive bioterrorism strategy, owing, in part, to the division of responsibilities between various federal departments and the subnational organizations (e.g., health authorities) that have primary responsibility for health emergencies (Public Health Agency of Canada, 2010). Generally, responsibility for bioterrorism is divided between Health Canada's Centre for Emergency Preparedness and Response (support for health-care providers) and the Canadian Food Inspection Agency (CFIA) Food Emergency Response System (responsible for identification and recall efforts) (Food Safety Network, 2008). The Public Health Agency of Canada is also responsible for bioterrorism, including planning, management, support, and scientific support (Public Health Agency of Canada, 2010). However, the Department of Agriculture is responsible for agriculture and agrifood in the event of a public safety emergency (including terrorism). In a terrorism-caused food safety crisis, the responsibility falls to the Royal Canadian Mounted Police in coordination with subnational and local police forces. Canada does not have a public food safety organization to combat bioterrorism that coordinates all of these organizations.

The French Food Safety Agency considers bioterrorism as part of its mandate, as identified in its 2004–2007 priorities (Évaluation des Risques Nutritionells et Sanitaires) (French Food Safety Authority, 2004). However, terrorism, food security, and biosecurity are not listed as focus issues on the agency's website nor are security or law enforcement agencies listed on its partnerships page. As a result, it seems that the French have not adopted a comprehensive or law enforcement-involved bioterrorism plan.

In 2002, Ireland commissioned the Expert Committee, Contingency Planning for Biological Threats, which created a framework policy to deal with bioterrorism from a health-care perspective (Government of Ireland, 2002). This report began by noting that: "It seems unlikely that a country such as Ireland would be the target for a primary bioterrorist attack" (Government of Ireland, 2002, p. 7).

The Japanese Food Safety Commission maintains procedures to cover bioterrorism. These policies are designed to facilitate *whole-of-government* responses (Japan Food Safety Commission, 2008). There is little specific information available.

Finland does not currently have a counterterrorism strategy; it undertook the creation of such a strategy on February 2, 2010 (Finland Ministry of the Interior, 2010).

Information on bioterrorism plans for Austria, Belgium, Denmark, Germany, Italy, Norway, the Netherlands, Sweden, Switzerland, if they exist, was not available.

Our risk society is no longer the same, or at the very least, in it is the transition. Our farming practices are addicted to chemicals. This may sound as a detriment to our environment by stating so. But in reality, chemicals allow to feed ourselves inexpensively, although we do not understand the real systemic costs. We consume food very differently than we did a decade ago and will do it differently in a decade from now. Biosecurity and bioterrorism go hand in hand. These particular topics have gained traction in the years since 2001. Since then, methods used to monitor and anticipate risks have changed. The way we perceived risks has changed as well, all around the world, not only for consumers but for farmers as well. Pesticide application is increasing around the world and despite extensive educational programs, farmers, who often expect higher yields, continue to take high health and environmental risks when applying pesticides. Expert opinion differs from farmers'. Some suggest this is due to differences in culture and tradition, the level of trust in the source of information, and the feedback on knowledge farmers are exposed to (Schoell and Binder, 2009). The food industry has not been immune to this significant shift over the last decade. We have seen changes in policy, which suggest governments and industry alike are mutating toward a different food safety surveillance model that often favors proactiveness versus reactiveness. Food safety has been at the forefront of newspapers and headlines, most specifically in the Western world, that it is increasingly difficult not to recognize its socioeconomic preponderance.

Biodiversity is a treasurable aim, which hangs in the balance when considering the environment and agricultural exploitation of resources. It is, indeed, an imperative natural capital asset that provides consumers and everyone profiting from natural resources with several goods and services that play a critical role in the economic and social wellbeing of humans in general. It also plays a central role in the evolution of ecosystems and enhances their resilience. Food safety measures related to biosecurity are particularly attractive because they can be used to increase the efficiency and cost-effectiveness of environmental management in a food safety context, create incentives for different investments related to land and crop management, create financial resources to support biodiversity conservation, and expand the scope of how the private sector is involved in environmental protection. Biosecurity inherently connects food systems with the land stewardship values. Food safety measures that have been used to conserve biodiversity include many different policies and do vary from one country to another. Successful food safety measures for this purpose requires the valuation of biodiversity and other biological resources, the involvement of all stakeholders, and the existence of appropriate policy and legal frameworks. Particular attention needs to be directed toward having an appropriate mix of instruments to harness any synergies and complementarities between them and to remove or mitigate perverse incentives. The experience from both developed and developing countries shows that if properly designed and implemented, economic instruments can significantly increase the returns to activities that conserve biological diversity and discourage behaviors that are detrimental to species and ecosystems. To see significant changes though, time is of essence. Consumers are to be educated on how agricultural chemicals can be a bearing to the environment and their health as well.

Some studies suggest that environmental friendliness appears to be inversely related to financial performance. The use of agricultural chemicals and land stewardship at a more holistic level remains highly immature. The corporate social responsibility has yet to influence or alter methods used in agriculture, at least for the time being. It is possible that due to the fresh emphasis and growing importance of environmental issues, organization in agricultural and agribusiness as a whole may not have had sufficient time to reap the full benefits of their decisions (McPeak et al., 2010). The biosecurity agenda is global. The developing economies of the world are feeding the rich. The interconnectivity between these worlds is creating disparities. The biosecurity metric may be skewed by globalization. Poor countries are able to produce at a lower cost over the short term, but long-term costs are significant but difficult to measure. The climate is not always favorable either. Chinese farmers, for example, have applied heavy doses of chemical fertilizer and pesticides to overcome natural resource constraints and significant pest pressures. The developing world's food industries have been stung by quality and safety problems both in overseas and domestic markets due to the pressures coming from rich countries. Having an impact on the production costs borne by farm households, applications of agricultural chemicals and exploitation of natural resources can have external costs borne by others or by future generations. To measure the use of agricultural chemicals per country in isolation is somewhat of a misleading process.

Therefore, more and more country will have to embrace environmental pluralism in the balance when monitoring risks.

Environmental pluralism becomes a source of environmental uncertainty when looking at food safety measures. Such a premise does certainly apply to the use of agricultural chemicals. It broadens the span of management for agrifood and implies that the macro environment consists of the industry's environment. This environment can be thought of in terms of proximity or distance from the food industry, with primary being the closest and macro being the farthest removed. In the case of possible food safety measures related to agricultural chemicals, agrifood's primary task environment would consist of relationships it has with both feed suppliers and consumers.

Concerning food safety policies and standards, agrifood has very little control over secondary and macro-task environments, and they are usually reactive to critical situations, for example, rather than proactive. Tasks derived from international markets are to be considered as macro-task environments. The scope of markets is not only limited to considering public policy but includes managing impacts of such policies. When considering agricultural chemicals, foreign markets can become the source of environmental uncertainty for agrifood industries included in the survey, a source that arguably was considered out of scope. Thus, agrifood depends on elucidations and ontological concepts offered by other markets, an approach that will encompass all three levels of task environments for any given food safety strategy.

When dealing with food safety issues, the world, through globalization, is doomed to undergo crisis after crisis. Agricultural chemicals may not lead to a crisis, but they present the potential to trigger trade disputes, protectionist measures among nations. Food safety measures were implemented across the world to safeguard the health of consumers, but policies have proved to unexpectedly fail all over the world in spite of these measures, increasing food safety concerns. These increased food safety concerns will create a new socioeconomic context in which production and consumption will be more harmonized. The food industry or agriculture itself is strategically inert and forced policymakers to concentrate mostly on short-term opportunities.

A more comprehensive approach to markets around food safety and the use of agricultural chemical practices would help agrifood appreciate the complexity of the macro-task environment surrounding their industry and how it could affect strategies in the future, particularly issues surrounding food safety. Such an approach would include risk management for industry in a food safety prone environment, and, most likely, ethics. Markets around the world are becoming more susceptible to food safety problems inherent in food trading activities between nations, which in return should force agribusinesses to enhance food safety responsiveness. We ought to suspect that agricultural chemicals may trigger more problems in the years to come.

For any given nation, regarding future crisis management procedures, the industry will have to accept as true that domestic and foreign food safety policies are slowly becoming one. Practices related to agricultural chemicals are slowly harmonizing but the process will likely take decades. This does not necessarily mean that all standards between nations will become one and the same. It is very unlikely that the world will ever apply homogenous food safety standards for contaminants, herbicides, and pesticides, as food safety policymaking is, in essence, a complex and somewhat diverse process. The industry would have to consider the most rigorous of standards amid targeted markets as being the model under which they should operate.

As for bioterrorism, consumers often forget. September 11, 2001, undoubtedly represents a historical date that will be forever printed on consumers' memories. The world saw significant changes in the outset. Homeland security was created in the United States and security measures at several borders were tightened, which to a certain extent distorted trade between countries. These events impacted how we perceived or managed risks may not be long lasting. The Second World War was devastating, but many do not recognize how devastating the war was to Europe and to other countries involved, including Canada (Milerski, 2010). Same rule could apply to food safety crises we have seen in recent years.

In Canada, the listeriosis outbreak in 2008 was a warning that could not be ignored. It reminded us how vulnerable consumers are to threats generated from our food supply. A recall led to the temporary closure of a Toronto-area processing plant, owned and operated by Maple Leaf. Although no link between the outbreak and the plant were found at the time, the company reacted vigorously and should be commended. Even so, Canadians have died despite the fact that, with the proper infrastructure, all food recalls are preventable. The problem is that our government-monitored food supply in its current state is no longer capable of protecting Canadians and therefore risks losing public trust.

It is chilling to read forecasts published in the last decade by food safety experts. Some analysts suggest the next 9/11 will occur through our food supplies (MacPherson and McConnell, 2007; Choe et al., 2009). Such a menace is particularly imaginable because our food safety architecture is arguably inadequate. It took 7 months to find the source of contamination in the 2006 American spinach recall. Even more worrisome, we found out that peppers were the culprits in the salmonella outbreak that hospitalized thousands in 2008 (Yasuda, 2010). Our ability to trace and track products in North America, let alone Canada, is recognized as being highly deficient (Starbird and Amanor-Boadu, 2006).

No individual organization is capable of meeting these challenges, not even trusted governmental authorities. In recent years, Canadians have had faith that government knows best when it comes to public health issues, and rightly so. Our public health system has served us so well for decades that it has become unnatural to think that profit-driven organizations care for the common good. But while many Canadians believe we should rely solely upon publicly funded authorities, the expanding scope of modern food systems is debunking such wishful thinking.

The food industry is a loose collective of organizations whose primary goal is to provide safe food to Canadian consumers, but its efforts are currently failing. Studies suggest that only a small percentage of everything Canadians eat is audited by competent public authorities (Hepner et al., 2004). Consider imports, restaurants, caterers, hot dog stands on city sidewalks, chocolate bars and chips bought in drug stores, or food purchased at events. The entire Canadian food industry represents well over $100-billion worth in annual revenues so the span of our food industry is enormous, but the majority of foods we eat are not screened at all. For the food industry to be capable of meeting its mandate, the private sector needs to play a proactive role with public agencies in food safety practices. Food safety authorities in this country need to build reliable partnerships to counter potential threats from the food supply, human induced or not. When considering bioterrorism or agriterrorism where an ill-intended group would try to harm a large group of consumers, our systems are nearly as adequate as they should. Supply chain security to mitigate the threat of agriterrorism falls to many organizations (and budgets) than just the producers.

The powers that be in food safety need to build reliable partnerships to counter potential threats coming from the food supply, whether these threats are human induced or not. To effectively manage risks, accountability, transparency, and responsiveness are key features we need to foster within food safety systems. With its crew of over 5000 people, one should not be too concerned about the CFIA's capacity to investigate in the event of an outbreak. Since its inception in 1997, the CFIA has matured into a valuable organization that is willing to learn from past experiences. Nonetheless, markets and Canadian consumers' behaviors have changed so rapidly. The CFIA and other provincial/municipal food safety authorities can hardly keep up and current public food safety resources are overexpanded. The CFIA's mandate should focus on establishing a consensus among food industry culprits so converging interests are shared.

We also need to redefine the geographical scope of current food safety systems. We now import well over $20-billion worth of food products from over 190 countries every year in Canada. Geographically, we may need to incorporate American authorities in the realm, at the very least. To set up a continental-based system would certainly be challenging, but it is necessary to proactively manage risks in the future and properly mitigate any agriterrorist threats.

When developing food risk management strategies, it is crucial to consider how consumers evaluate food risk practices. Proactive consumer protection, for example, is often positively related to consumer evaluation of food risk management quality. Proactive measures include enhanced food traceability, education and awareness, surveillance, proper risk management certifications, and improved supply chain control. These responsibilities should be shared between the public and private sectors. Closer cooperation will help identify underlying problems and anticipate future threats.

We are faced with a threat, but also an opportunity to improve Canadian food safety. Rather than forcing governmental authorities to play the role of industry enforcer, we must protect the rapport between Canadian consumers and the food industry before it is too late.

5

Governance and Recalls

The evaluation of Governance and Recalls is akin to the macrolevel analysis of biosecurity risks. Just as biosecurity concerns—either the application of a chemical pesticide or the risk of terrorism—can impact the perception of food safety in the modern ecologically sensitive and terrorism-concerned world, the study of Governance and Recalls is concerned with the preventative and remediative issues facing food safety regulators, particularly as they concern producers.

It is clear from Chapter 3 that food regulators and food consumers form an important stakeholder relationship. What is important also to remember, despite the obvious interest divergences and power differentials, is the interaction between the agrifood industry and government-run food safety regulators. In particular, how this impacts the overall level of food safety within a jurisdiction. Whether through the adoption of Hazard Analysis and Critical Control Points (HACCP) safety systems, shifting international best practice norms, or the external pressures caused by foreign trade, the global food system has an impact on domestic producers, large and small. While some of the pressures on producers (or any business) may be changing because of globalization, the ultimate direct arbiters of food safety are domestic authorities (even, in the case of the European Union (EU), where some authority rests at a supranational level).

Assuming that these global pressures exist and that they impact modern, open countries (like the 17 studied in the *Food Safety Performance World Ranking Initiative*), what tools exist for food safety regulators to implement best practices? Presumably, some of these can be measured by outcome. Just as one can measure the level of chemical pesticide use as a byproduct of a country's specific need and its level of biosecurity concern, so too are national risk management plans a reflective outcome of its food safety concerns. So, are these plans clear and effective? Furthermore, in addition to studying a concrete outcome like risk management plans, do a country's national policies help by providing teeth to

Food Safety, Risk Intelligence and Benchmarking, First Edition. Sylvain Charlebois.

its risk-minimization efforts? For example, when something goes wrong, how does the authority fix it? How often is this corrective system triggered? How sensitive and responsive is it?

Ultimately, these policies and the observed outcome are a measurement of and a reflection upon a nation's food safety regulators. However, these policies and outcome rest, in part, on the active participation and support of producers. Assuming that a producer would never want to intentionally harm its customers and a regulator's sole interest is safe food, then both have a mutually beneficial opportunity to collaborate on the establishment and implementation of food safety best practice. Naturally, in the real world, each party has its own self-interests, so debate—at some level—between the regulator and regulated are likely inevitable. However, even if idealized cooperation may be unlikely, the closeness of the relationship results in a somewhat symbiotic bond between the governance and recall efforts at the national, food regulator level, and its impact on producers.

In the short run, agribusiness producers fear the imposition of new regulation (Sapp et al., 2009). Whether this concern is because of selfish profit motivation, a legitimate reaction to an overzealous regulator trying to expand its authority, or a dispute about standards or costs, the fear of rapidly changing regulations is understandable. While few (save, possibly, some bureaucrats with their own selfish interests at heart) would welcome dramatic shifts in policy without due cause, when food safety regulators develop and implement best practice on a national level, these policy changes can have positive domestic and international results for producers.

Initially, the stability of well-established and effective food safety rules represents a vote of confidence in the *status quo* on the part of all stakeholders. Moreover, it also allows producers to work within familiar standards and to contemplate business investments with confidence in the regulatory framework down the road. Product safety is a logical concern for a producer and the recognition of effective domestic food safety rules a follow-on benefit. Just as this is the case domestically—for, after all, what consumer would prefer to frequent an unsafe or unregulated producer—the same holds true at the international level.

In addition to reviewing international risk management and food recall-related best practices, the *Food Safety Performance World Ranking Initiative* also considers a country's trading policy. Using a country's simple average most favored nation (MFN) rate as a barometer of protectionism, one observes the bluntest possible measure of how a country feels about its trading partners' agricultural products. A high protectionist rating could reflect a country's concern about the safety of imported food. If the opposite were true, countries without strong protectionist policies are demonstrating confidence in food safety best practice policies—the adoption of which would then be allowing producers to move into new international markets.

The adoption of food safety practices by producers falls outside of the scope of the *Food Safety Performance World Ranking Initiative.* The State–Pressure–Response model used to evaluate individual countries' food safety protocols is based on using international best practice in the context of a country's specific needs. Nevertheless, the relationship between regulator and producer as a result of the State–Pressure–Response model and the best practice outcomes and policies exists at the heart of the Governance and Recall study.

Governance and Recalls in the *Food Safety Performance World Ranking Initiative*

The investigation into Governance and Recalls comprised one outcome and three policies; respectively, the existence of risk management plans, the level of clarity and stability of food recall regulations, and the number of protectionist measures against trading partners and the number of recalls. Countries were ranked based on their performance in these criteria on a sliding scale from superior to poor and then graded against their peers.

In addition to the four criteria, country-specific information was considered. For example, Japan imports about 60% of its food (Japan Ministry of Health, Labour and Welfare, 2010); therefore, its traceability and management systems should be different from other countries. Moreover, while the absolute number of recalls inside a country is an important metric, the category was also considered in the context of absolute population size (one would expect larger countries to have more types of food products and larger food industries, which are likely to increase the total number of voluntary recalls regardless of the vigilance of local inspectors).

Australia, Canada, Japan, and the United States (US) each received grades of superior. The average-graded countries, Austria, Belgium, Denmark, Italy, the Netherlands, and the United Kingdom (UK), each had four criteria grades of moderate. Norway, which also earned a grade of average, had two criteria grades of moderate. Finland, France, Germany, Ireland, Sweden, and Switzerland all earned poor grades.

Comparison with 2008

As illustrated in Table 5.1, there were large changes in category grades and rankings from 2008 to 2010. Australia, Austria, Belgium, Italy, and Japan all improved their grades in 2010 compared to 2008. In most cases,

Table 5.1 Ranking for governance and recalls.

			2008 Comparison	
Rank	**Country**	**Grade**	**Rank**	**Grade**
1	Canada	Superior	3	Superior
2	Australia	Superior	7	Average
3	United States	Superior	2	Superior
4	Japan	Superior	6	Average
5	Netherlands	Average	1	Superior
6	United Kingdom	Average	4	Superior
7	Belgium	Average	12	Poor
8	Denmark	Average	8	Average
9	Austria	Average	11	Poor
10	Italy	Average	17	Poor
10	Norway	Average	5	Average
12	Germany	Poor	15	Poor
13	France	Poor	14	Poor
14	Ireland	Poor	9	Average
15	Finland	Poor	13	Poor
15	Sweden	Poor	16	Poor
15	Switzerland	Poor	10	Average

Increased grade	No change	Decreased grade	No data

this was in part thanks to decreases in agriculture MFN rates. Belgium also managed to increase three of its four criteria grades.

Ireland, the Netherlands, Switzerland, and the United Kingdom all received lower grades in 2010 than 2008. In the case of the United Kingdom, it fell in three of the four criteria grades—earning four moderates.

Existence of Risk Management Plans

Analysis of Ranking Data

Risk management is an aspect of the risk analysis process undertaken to set policies for food safety. The CODEX standard for risk management procedures includes undertaking preliminary risk management action, evaluating options, implementing and monitoring a risk management policy, and evaluating the outcome of the decision.

The importance of a risk management plan lies in the evaluation and implementation of food safety policy. In other words, a well-functioning risk management system will have preventative and corrective actions. However, the clarity of these plans is also an essential factor in their success. Because none of the countries included in this survey rely on its food safety organs to produce food products, the clarity of the food safety organization's tools is important so that producers understand what is required to prevent, monitor, and solve any food safety concerns.

This metric considered whether a risk management plan for food safety is clear. The various plans were graded based on the clarity of who is making the policies to ensure safe food and on what basis they make their decisions. In effect, this metric evaluates who is making the food safety (risk management) policies and what factors inform that decision. The greater the clarity from the food safety system, the higher the country's grade. Countries with clear, well-established, thorough and science-based risk management systems earned grades of progressive. Risk management systems with gaps or only semiclear objectives earned their countries grades of moderate. Countries without clear or well-defined risk management systems were given grades of regressive.

Australia, Canada, and Japan all earned grades of progressive. They each have clear, thorough and science-based risk management systems. The EU countries (Austria, Belgium, Denmark, Finland, France, Germany, Ireland, Italy, the Netherlands, Sweden, and the United Kingdom) earned grades of moderate because EU rules allow for individual countries to interject nonscience-based judgments into risk management policies. Switzerland does not provide information on its risk management plans; therefore, it earned a regressive grade. So too did the United States, which also does not provide clear information on its risk management systems.

Comparison with 2008

As illustrated in Table 5.2, Ireland was the only country to improve on its 2008 grade in 2010. Like the rest of its EU partners, the Irish have a generally adequate risk management system, which is in keeping with the CODEX standard. However, like its moderately graded counterparts, its risk management system allows for the possibility that hidden or irrelevant policy considerations could impact risk management decision making.

Like Ireland, France and the UK's risk management systems fall victim to the risk of unclear policy setting. Because the EU-required risk management program allows too much variability in risk management policy setting, France and the UK's systems were not considered progressive grade.

Table 5.2 Existence of risk management plans.

Country	2008 Grade	2010 Grade
Australia	Progressive	Progressive
Austria	N/A	Moderate
Belgium	N/A	Moderate
Canada	Progressive	Progressive
Denmark	N/A	Moderate
Finland	N/A	Moderate
France	Progressive	Moderate
Germany	N/A	Moderate
Ireland	Regressive	Moderate
Italy	N/A	Moderate
Japan	Progressive	Progressive
Netherlands	N/A	Moderate
Norway	N/A	Moderate
Sweden	N/A	Moderate
Switzerland	N/A	Regressive
United Kingdom	Progressive	Moderate
United States	N/A	Regressive

Increased grade	No change	Decreased grade	No data

Level of Clarity and Stability of Food Recall Regulations

For all of the same reasons that the level of risk management clarity is important, so too is the clarity and stability of a country's food recall regulations. By providing a constant and clear set of rules and procedures, food safety authorities allow producers to understand their responsibilities. At the same time, the food safety authorities have the ability to fine-tune already existing programs, allowing for continuity of policy administration.

Countries that provided clear information on a food recall policy earned grades of moderate. To earn a grade of progressive, a country must not only have clear and stable food recall rules but also provide follow-up monitoring and enforcement of a recall. Countries without a clear or complete food recall system earned grades of regressive.

Canada and the United States both earned grades of progressive. They each have clear and comprehensive food recall regulations, while also providing significant information on their follow-up procedures. Australia, Austria, Belgium, Denmark, Finland, France, Germany, Ireland, Italy, the Netherlands, Norway, Sweden, Switzerland, and the United Kingdom all earned grades of moderate. In the case of Australia, its food recall system is somewhat confused with its subnational governments. The European countries have adequate food recall regulations (set by the European Union but disseminated by a pan-European organization); however, the legislation does not require follow-ups. Japan earned a grade of regressive for a lack of data on any domestic food recall system.

Comparison with 2008

As illustrated in Table 5.3, the only countries to improve their scores were either part of the European Union or (in the case of Norway) adopted

Table 5.3 Level of clarity and stability of food recall regulations: 2008 and 2010 grades.

Country	2008 Grade	2010 Grade
Australia	Progressive	Moderate
Austria	Regressive	Moderate
Belgium	Regressive	Moderate
Canada	Progressive	Progressive
Denmark	Regressive	Moderate
Finland	Moderate	Moderate
France	Moderate	Moderate
Germany	Moderate	Moderate
Ireland	Moderate	Moderate
Italy	Moderate	Moderate
Japan	Moderate	Regressive
Netherlands	Moderate	Moderate
Norway	Regressive	Moderate
Sweden	Regressive	Moderate
Switzerland	Moderate	Moderate
United Kingdom	Progressive	Moderate
United States	Progressive	Progressive

Increased grade	No change	Decreased grade	No data

EU rules. The improvement of Austria, Belgium, Denmark, Norway, and Sweden reflects this food recall standardization.

Australia, Japan, and the United Kingdom each scored lower in 2010 than in 2008. In the United Kingdom, this decline reflected the standardizing process that raised the grades of the other European countries. Australia fell because of confusion in its food recall data, which is not in keeping with the progressive-graded countries. Therefore, its system is comparatively moderate. Japan fell in 2010 because of the absence of data on its domestic food recall system. As no comprehensive data can be found, its food recall system cannot be considered clear.

Number of Protectionist Measures Against Trading Partners

This measure can be controversial, but the council of academics deemed trades to enhanced risks domestically, which is why this category was added to surveys at the time. Mechanisms like GFSI were not as influential but progress on this matter is reflected in the 2014 survey presented at the end of this book.

All of the 17 countries in this study are members of the World Trade Organization (WTO), which requires that each member country treat the other member countries' exports with the same tariff rates. These are known as MFN rates. This is supposed to harmonize the tariff rates that a country applies to goods imported from all WTO members. However, the MFN rate is effectively a price ceiling because countries can make non-WTO agreements (like a free-trade deal) with other countries to provide better than MFN rates (World Trade Organization, 2010a). Therefore, the higher a country's MFN rate (at least with other WTO members), the higher the tariffs that it is charging. (Because countries are required to offer MFN or better, a higher MFN rate means fewer lower-tariff agreements.)

The simple average MFN rate applied to agriculture is the average MFN tariff rate that the country charges to foreign countries trying to export agricultural goods to the destination country. As a result, the higher a country's MFN rate for agricultural goods, the greater the tariff barrier against foreign agricultural products. A higher MFN rate could have the effect of protecting a domestic industry from foreign competition. However, a higher MFN rate could also be a sign that a country has concerns about its food safety inspection procedures for imported food or that the country has concerns about the food safety of its trading partners. Conversely, the lower a country's MFN rate on agricultural products, the more that it trusts trading partners' food safety and its own detection, recall, risk management, and general food safety apparatuses.

Countries with agriculture MFN rates of less than 10% earned grades of progressive. Countries with MFN rates between 10 and 25% earned grades of moderate. To earn a grade of regressive, a country's agriculture MFN rate must be above 25%.

Australia and the United States were the only countries to earn progressive grades. Their agriculture MFN rates are only 1.3 and 5.3, respectively. Canada, the EU countries, and Japan all earned grades of moderate for their rates of 11.5, 16.0, and 23.6, respectively. Norway and Switzerland both have high agriculture MFN rates at 59.0 and 44.0, respectively.

Comparison with 2008

As illustrated in Table 5.4, the European Union saw an improvement in 2010. Because of the common market, all EU countries have the same external tariff rates on agricultural goods. Since 2008, the Union's

Table 5.4 Number of protectionist measures against trading partners: 2008 and 2010 grades.

Country	2008 Grade	2010 Grade
Australia	Progressive	Progressive
Austria	Regressive	Moderate
Belgium	Regressive	Moderate
Canada	Moderate	Moderate
Denmark	Regressive	Moderate
Finland	Regressive	Moderate
France	Regressive	Moderate
Germany	Regressive	Moderate
Ireland	Regressive	Moderate
Italy	Regressive	Moderate
Japan	Moderate	Moderate
Netherlands	Regressive	Moderate
Norway	Regressive	Regressive
Sweden	Regressive	Moderate
Switzerland	Regressive	Regressive
United Kingdom	Regressive	Moderate
United States	Regressive	Progressive

Increased grade	No change	Decreased grade	No data

comparable position has improved, resulting in an increase in grade from regressive to moderate. The United States also lowered its MFN rate, jumping two grades from regressive to progressive. The United States has the second lowest (to Australia) agriculture MFN rate among the 17 countries in this study.

No country scored lower in 2010 than in 2008. Because the agriculture MFN rates required for each grade remained the same as 2008 (Charlebois and Yost, 2008), tariff barriers set by the countries in this study have fallen.

Number of Recalls

Perhaps considered as another controversial metric, the number of recalls was added on the basis that our council requested it. As one of its frequently asked questions, the US Food and Drug Administration (FDA) answers: "Why do we still have food recalls and outbreaks of food-related illness?" Part of that organization's answer is: "Unpredictable events, mechanical and human error, and environmental conditions all play a role in the problems we continue to see in food production, processing, and distribution." The reality is that food safety issues are unlikely to disappear. In fact, this is why food safety authorities are so important and why it is necessary to acknowledge the relative performance of each country's system (this assumption underlies the purpose of this study).

Food recalls form an important aspect of the responsibilities of food safety organizations. As such, this metric evaluates the number of food recalls in each country, whereby the best-ranked countries record the highest number of food recalls. Because food safety issues continue to crop up, in effect the number of food recalls in a given period is evidence of the strength of a country's food safety system. As explained by the EU Rapid Alert System for Food and Feed (RASFF) in its 2008 annual report:

> Some caution needs to be exercised when drawing conclusions from [the number of notifications, the origin of the notifications, the countries involved, the products and the identified risks]. For example, it is not because a Member State has a relatively high number of notifications that the situation regarding food safety would be bad in that country. On the contrary, it could indicate that a greater number of food checks are carried out or that the communication systems in that Member State function well.

This comment applies to both the 2008 and 2010 surveys, but the approach changed in 2014 as it is explained in the last chapter. Countries with over 50 food recalls were considered progressive. Those with between 50 and 10 recalls were moderate, and those with fewer than 10 food recalls were given grades of regressive. Australia, Canada, Germany, and the United States all earned grades of progressive. Austria, Belgium, Denmark, France, Italy, the Netherlands, and the United Kingdom earned grades of moderate. Finland, Ireland, Norway, Sweden, and Switzerland all earned grades of regressive. Japan, without any data, was given a grade of not applicable.

Comparison with 2008

As illustrated in Table 5.5, Belgium and the United States both increased their 2010 grades compared to 2008. The United Kingdom, which earned a grade of progressive in 2008, fell one food recall (49 out of 50) below the minimum required for a progressive grade. As a result, it fell from progressive to moderate in 2010.

Table 5.5 Number of recalls: 2008 and 2010 grades.

Country	2008 Grade	2010 Grade
Australia	Progressive	Progressive
Austria	N/A	Moderate
Belgium	Regressive	Moderate
Canada	Progressive	Progressive
Denmark	N/A	Moderate
Finland	Regressive	Regressive
France	N/A	Moderate
Germany	N/A	Progressive
Ireland	Regressive	Regressive
Italy	N/A	Moderate
Japan	Moderate	N/A
Netherlands	N/A	Moderate
Norway	N/A	Regressive
Sweden	N/A	Regressive
Switzerland	N/A	Regressive
United Kingdom	Progressive	Moderate
United States	Moderate	Progressive

Increased grade	No change	Decreased grade	No data

Investigation on Governance and Recalls

The investigation on Governance and Recalls comprised four sections: one outcome and three policies. Their measurement informed the ranking for Governance and Recalls.

Existence of Risk Management Plans

When considering each country's risk management system, this study considered whether the responsible authority (or authorities) was undertaking preliminary risk management actions, evaluating their options, implementing and monitoring the risk management policy, and evaluating their policy's outcome.

Australia has a very clear risk management system, with food standards set by the Food Standards Australia New Zealand (FSANZ) agency. Its risk analysis process "[examines and incorporates a] wide variety of factors that impact on a decision-making process." Risk management, in Australia, is a four-step process: an examination of the nature and potential impact of the issue, establishment of a broad risk management goal and steps to achieve them, consideration of possible options and making a decision, and the implementation of the risk management control (Food Standards Australia New Zealand, 2009). Moreover, the FSANZ also clearly lists the factors that influence its risk management decisions:

> In developing risk management decisions, FSANZ must consider the objectives of the Authority... These objectives include:
>
> i) the protection of public health and safety;
> ii) the provision of adequate information relating to food to enable consumers to make informed choices; and
> iii) the prevention of misleading or deceptive conduct.
>
> In addition to these objectives FSANZ must also have regard to:
>
> i) the need for standards to be based on risk analysis using the best available scientific evidence;
> ii) the promotion of consistency between domestic and international food standards;
> iii) the desirability of an efficient and internationally competitive food industry;
> iv) the promotion of fair trading in food; and
> v) any written policy guidelines formulated by the Council for purposes of this paragraph and notified to the Authority.

> In considering these objectives, FSANZ takes into account a number of different issues including human health, consumer behaviour, economic, governmental and international agreements. (Food Standards Australia New Zealand, 2009)

The FSANZ also lays out its options for managing the actionable issues identified by its risk management assessment (Food Standards Australia New Zealand, 2009).

In Canada, the Canadian Food Inspection Agency (CFIA) has established enhancements to its risk management integration as one of its top five priorities. Moreover, it has identified nine "key strategic risks": Foodborne hazards; zoonotic outbreaks/incidents; animal and plant pest hazards; science and technology capacity; program framework; partnerships: roles and responsibilities; human resources capacity and capabilities; data, information, and knowledge for decision making; and internal coordination. Furthermore, the CFIA has created multiple strategies to resolve each individual risk. While only the first two risks deal specifically with making clear, science-based food safety risk management decisions, the CFIA has further created policies and risk management strategies to combat organizational dangers—for example, those that could endanger the CFIA's ability to accomplish its responsibilities (Canadian Food Inspection Agency, 2010c).

The European Union has a community-wide rule governing risk management plans:

> Risk management shall take into account the results of risk assessment, and in particular, the opinions of the [European Food Safety Authority (EFSA)], other factors legitimate to the matter under consideration and the precautionary principle [where relevant], in order to achieve the general objectives of ... food law. (European Parliament and European Council, 2002)

Like the EU principles of risk assessment, EFSA opinions are based on risks and scientific study (European Parliament and European Council, 2002). However, the European Union also allows for "consideration [of] a wide range of other factors legitimate to the matter under consideration" by decision makers. The possibility that these other criteria may cloud the ultimate risk management decision—especially because the administration of these decisions is ultimately left to the member country—is a negative mark against the EU system.

Japan has a bifurcated system of risk management: The Japanese Ministry of Health, Labour and Welfare (MoHLW) is responsible for food sanitation risk management, while the Japanese Ministry of Agriculture,

Forestry and Fisheries (MoAFF) is responsible for agricultural, forestry, and fisheries product risk management (The JFSC is responsible for risk assessment and communication.) (Japan Food Safety Commission, 2010). The MoHLW specifically outlines the responsibilities and flow of decision-making powers (Japan Ministry of Health, Labour and Welfare, 2009). However, it does not clearly explain the decision-making process. On the other hand, the MoAFF does have clear principles for its risk management principles and implementation strategies (Japan Ministry of Agriculture, Forestry and Fisheries, 2010).

In 2004, Norway reorganized its food safety organs, which included the chartering of the Norwegian Food Safety Authority (NFSA). This body takes the science-based risk assessments generated by other divisions and implements them in Norway. Other than a general description of the scientific resources used to generate the risk assessments, there is little detail on the risk management policies or procedures used by the NFSA.

Switzerland does not provide information on its risk management plans. (However, it is negotiating with the European Union to create cooperative measures on food safety, including risk management and participation in the EFSA (Switzerland Federal Office of Public Health (FOPH), 2010).)

In the United States, food safety is a shared responsibility between the FDA and the Department of Agriculture (USDA). However, neither organization clearly lays out its risk management considerations or policymaking process. Therefore, the US's risk management system could not be considered clear nor could policy-decision rationales be evaluated.

Level of Clarity and Stability of Food Recall Regulations

This study considered whether clear information was available on food recall regulations in each of the 17 countries. Those countries that also provided follow-up oversight earned higher grades.

In Australia, food recalls are managed and controlled by FSANZ, which has the power to require product recalls; coordinate responsibilities between itself, the involved industry, and the state governments; communicate recall-related information, and follow up on the effectiveness of the recall (Food Standards Australia New Zealand, 2008). Based on this information, Australia could be considered a progressive-graded country. However, food recall responsibility also rests at subnational levels. While coordination exists, there is a muddied set of responsibility and sometimes-conflicting information.[1] As a result, Australia fell from progressive in 2008 to moderate in 2010.

---

1 See the Australia discussion in number of recalls.

In Canada, similar to Australia, food recalls are administered by a centralized food safety agency: the CFIA. Its responsibilities include investigating food safety issues, issuing notice of recalls, ensuring that *corrective measures* are taken and ensuring that the food safety issue was resolved when production resumes (Canadian Food Inspection Agency, 2010b). Canadian food recalls (most of which are voluntary—although the CFIA has the power to issue mandatory recalls) fall into one of three categories: Class I (high risk), Class II (moderate risk), or Class III (low and no risk) (Canadian Food Inspection Agency, 2010a). In 2008–2009, the CFIA met its target of providing public warnings of Class I recalls within 24 h of a decision. This is the third year in a row that the agency has met this target (Canadian Food Inspection Agency, 2010b). Because the CFIA provides above-average food recall services and is clear about its requirements and procedures, Canada earned a progressive grade in 2010.

The European food recall system is the average model against which all other countries are compared. The European Union has set down rules governing recall processes in the Union, which include the responsibility to recall food that does not comply with safety requirements and the notification of involved stakeholders (including most of continental Europe through the RASFF rapid-alert system. This system, which recently celebrated its 30th anniversary, provides European food safety agencies (including the 13 European countries covered by this study) with recall information and centralizes outcomes. There is nothing missing or unstable about this system; however, it does not provide as many services as the progressive-graded countries' systems. They are, nevertheless, adequate.

There is limited information available on the Japanese recall system. What information is available concerns the rules and regulations regarding recalls of imported foods. Generally, this requires the adherence to local and foreign recall information, which occurs during the importation process. Local authorities are also required to undertake recalls if an imported product is found to be unsuitable after it has passed the inspection (Japan Director of the Department of Food Safety, Pharmaceutical and Food Safety Bureau, 2010). Because of incomplete information or an incomplete system, the Japanese food recall system cannot be considered clear or comprehensive.

Like Canada, the United States has an excellent food recall system. Food recalls are administered by either the USDA's Food Safety and Inspection Service (FSIS) or the FDA. The FSIS deals with meat, poultry and egg products, while the FDA takes care of all other food products. Both agencies have the power to enforce product recalls; however, generally, recalls are undertaken voluntarily by industry. Both recalling agencies undertake similar procedures: A recall is initiated, classified, the public is notified, the recall is monitored and, finally, the recall is terminated (United States

Department of Agriculture: Food Safety and Inspection Service, 2008; United States Food and Drug Administration, 2010b).

Number of Protectionist Measures Against Trading Partners

The data for agriculture MFN rates came from the WTO's 2009 *World Tariff Profile*. The relevant data are provided in Table 5.6.

Number of Recalls

The measurement of the number of food recalls in a country was based on the information provided by the relevant agency (of agencies) responsible in a given jurisdiction.

In Europe, food recalls are monitored and measured by the EU RASFF. It measures food safety notifications as alerts (serious risks, including recalls), information notifications (nonurgent issues), border rejections (when an import product is refused entry), or news notifications (information judged to be of interest). Table 5.7 provides the number of alerts in 2008 by RASFF member.

Table 5.6 Simple average MFN applied to agriculture in 2008.

Country	MFN rate	Country	MFN rate	Country	MFN rate
Australia	1.3	Japan	23.6	Switzerland	44.0
Canada	11.5	Norway	59.0	United States	5.3
European Union[a]	16.0				

Source: World Trade Organization, 2010b. World Tariff Profiles 2009. WTO Secretariat: Geneva, Switzerland. Page 34.

a) As the European Union comprises a common market, each member country has the same MFN rate.

Table 5.7 Number of alerts in 2008 by European country.

Country	Alerts	Country	Alerts	Country	Alerts
Austria	23	France	42	Netherlands	40
Belgium	41	Germany	102	Norway	3
Denmark	24	Ireland	9	Sweden	4
Finland	9	Italy	70	United Kingdom	49

Source: European Commission, Directorate-General for Health and Consumer Protection. 2009. The Rapid Alert System for Food and Feed (RASFF): Annual Report 2008. European Commission, Directorate-General for Health and Consumer Protection: Luxembourg.

Switzerland is a member of the EU RASFF; however, based on the RASFF's 2008 annual report, it did not issue any food safety alerts (including food recalls) in 2008.

In Australia, the FSANZ is the national body responsible for food recalls. According to the 2008–2009 FSANZ annual report, the FSANZ only responded to three incidents in that 12-month period. However, it is important to note that responsibility for food safety and recalls are shared between the FSANZ and Australia's substate actors (Food Standards Australia New Zealand, 2010). The FSANZ provides historical information on the number of Australian food recalls—up to and including 2007. Based on the previous 5-year average, Australia had 67.6 food recalls per year. Because of the large disparity between this 5-year average and the figure reported in the FSANZ annual report, and in further recognition of the distinction between FSANZ-coordinated recalls and those smaller recalls identified and handled at the substate level, the 67.6 figure was used for this metric.

Compared with Australia, Canada handles far more food recalls each year: generally about 350 (Canadian Food Inspection Agency, 2010a). In 2008, this included "one of the largest in Canadian history" when 192 products were recalled in the *Listeria* outbreak (Canadian Food Inspection Agency, 2010b).

Like Australia and Canada, in the United States recall responsibilities are divided. The US FSIS oversaw over 50 meat, poultry, and egg product recalls in 2009 (United States Department of Agriculture, 2010), while the US FDA oversaw hundreds (United States Food and Drug Administration, 2010a).

There are no data about the number of recalls in Japan in 2009 or other recent years.

Discussion

The relationship between a country's food safety agencies and the food industry is critically important. The 17 countries in the *Food Safety Performance World Ranking Initiative* are advanced capitalist democracies with modern and mature food industries that produce for domestic and international consumption. In each of these countries, large- and small-scale producers hold market power, which is regulated by the firm hand of each county's food safety authority. The importance of private industry is best understood in the context of the regulator–consumer relationship explored earlier (see Chapter 3). These two relationships, industry–regulator and regulator–consumer, form two-thirds of the holistic food sector. The third relationship, forming the industry foundation, is the producer–consumer relationship.

Food safety authorities around the world recognize the importance of these three participants and their relationships, particularly as they concern burden sharing. At the Second Global Forum of Food Safety Regulators held by the United Nation's Food and Agriculture Organization (FAO) and the World Health Organization (WHO) in 2004, Alan Reilly, then the Deputy Chief Executive of the Food Safety Authority of Ireland, highlighted the nature of the burden sharing. While regulators must work on behalf of the consumer, he said, the consumer must also assume some responsibility. Moreover, the effectiveness of government controls plays a role in assuring the safe production of food (Food and Agriculture Organization of the United Nations and the World Health Organization, 2004).

The power differential between the three stakeholder groups underlies the importance of burden sharing and responsibility. As discussed in Chapter 3, consumers are ultimately superior to both industry and government: They exercise market power by rewarding good producers and exercises political power in demanding regulatory change through elected representatives. However, there is a clear power differential when comparing the knowledge and direct market or electoral power of a single consumer with the expertise and resources commended by a regulatory or business organization.

Each of the 17 countries in the *Food Safety Performance World Ranking Initiative* organizes its electoral and regulatory systems differently. This introduces variables into the regulator–consumer relationship. To illustrate this point, consider Canada. Its chief food safety organization—the CFIA—is responsible to the federal government and currently in the portfolio of the Minister of Agriculture and Agri-Food and Minister for the Canadian Wheat Board Gerry Ritz. While accountable through the federal cabinet, Mr. Ritz is not electorally accountable to all Canadian consumers, merely those in his Battlefords–Lloydminister riding in Saskatchewan. Moreover, come election time, those constituents make their voting decisions based on any number of possible criteria, not necessarily the performance of the CFIA or agriculture. By comparison, in some regions of the United States, agriculture and food safety are ballot box issues through the popular election of food safety-responsible executives. For example, after an outbreak of Salmonella traced to two Iowa egg farms, the elected Iowa Agriculture Secretary faced a direct political challenge over the issue (Crumb, 2010). The Canadian example can be seen as a diffuse connection between consumer and regulator, while the American example represents a closer connection.

While each country has different specific relationships between their food stakeholders, the Government of Canada has presented to the FAO and the WHO that food production should be a "collaboration of all stakeholders" (Government of Canada, 2004). Nevertheless, how ever much the industry–regulator–consumer relationship is cooperative, Canada placed more emphasis on the industry–regulator relationship. Specifically:

> Government has the primary responsibility for identifying and assessing health risks associated with the food supply, and developing national strategies to manage the risks. Industry has the primary responsibility for the safety of its products and for providing appropriate information to permit consumers to make informed choices (Government of Canada, 2004).

Whether the Government of Canada's position is a result of the weaker power of individual consumers facing powerful producer or producer organizations, a recognition that Canada has a weaker political regulator–consumer relationship, or something else entirely, it underlines the critical importance of the industry–regulator relationship.

While there is obvious import that food safety regulators require the food industry to produce safe food, the *Food Safety Performance World Ranking Initiative* evaluates the success from the producer's point of view. In fact, the questions are: Can industry understand the rules (are risk management policies clear); How clear and thorough are recall rules (clarity, stability and follow-up of food recall regulations) and do they work (number of food recalls); and, How protectionist are national governments to imported food (simple average MFN rate)?

Because the food industry directly touches all citizens' lives daily, governments place enormous responsibility in the hands of industry to ensure that their citizens have access to safe nutrition. Despite the enormous importance of this responsibility, large and small agricultural producers must be able to fulfill this responsibility efficiently and in keeping with their business models. Therefore, business has a direct and obvious interest in knowing what agency is making the rules and how the rules are created. Moreover, in keeping with the biological and chemical risks, special care should be paid that food safety agencies make science-based decisions, rather than involving irrelevant considerations. Similarly, clear, stable and effective recall regulations—with checks to ensure compliance—present important information to industry.

These first three criteria have an obvious impact on the producer. In order to effectively monitor, manage and resolve food safety issues, national regulators require the direct participation of industry in managing risks and accomplishing product recalls. The fourth criterion, trade, is perhaps one of the most nuanced aspects of the *Food Safety Performance World Ranking Initiative*, especially as it concerns industry. International agriculture trade is growing between developed as well as developing countries (Canadian Food Inspection Agency, 2010d; Kolesnikov-Jessop, 2010; Produce Safety Project, 2010). This increase should represent an opportunity for large and small agribusiness, as new markets can mean new opportunities for growth and expansion.

In tandem with the increases in agritrade is a trend toward nondomestic recognition of foreign food safety policies. Although the countries are not included in the 17-nation *Food Safety Performance World Ranking Initiative*, the case of Singapore and China is instructive. Because of its geographical constraints, Singapore imports over 90% of its food. To help maintain a diverse food supply network, producers from the island city-state have turned to China as a new, large-scale source of food. Through special arrangements, Singapore companies are partnering with local Chinese governments to create new agricultural opportunities. What is particularly important about these partnerships is that the Singapore food regulator, the Agri-Food & Veterinary Authority (AVA), is also taking part (Kolesnikov-Jessop, 2010). In effect, Singapore producers intend to grow food in China to import into Singapore using AVA-regulated and inspected farms.

Without digging too deeply into a comparison between food safety in Singapore in China, this example demonstrates the global nature of modern food safety regulations. In this case, Singapore's AVA is importing itself into a foreign market to oversee best practice (in a Singapore context) for food that will then be imported. Although the full effect of this project will not be known for years (construction started on one project in September, 2010, but will take up to 6 years to reach its full potential (Kolesnikov-Jessop, 2010)), the importation of foreign food products is a hallmark of a country's modernization process (Cassels, 2006). It is also an accelerating trend in North America.

Between 1997 and 2006, Canada imported $21.8 billion (Canadian) of inflation-adjusted food value—a 45% increase from the $14.2 billion (Canadian) imported in 1997. Moreover, Canada currently imports from virtually every recognized country in the world. At the same time, the CFIA is placing increasing reliance on foreign-country food safety regulations. As indicated by a 2010 audit of the CFIA's import controls, Canada's inspection agency uses foreign-country equivalency assessments as a matter of policy but has yet to conduct formal evaluations for all countries. Furthermore, these controls only exist for meat, seafood, and eggs: "Imports of other food commodities rely almost exclusively on destination inspections and projects." The increased reliance on the exporting country's food safety agency is a deliberated CFIA goal to "reduce dependence on downstream [CFIA] controls" (Canadian Food Inspection Agency, 2010d). Regardless of the reason, the importance of external food safety agencies is a concern for a domestic food safety standard—at least when that country imports significant amounts of food.

Unlike Singapore, Canada is moving toward something of a food safety passport system, with the CFIA increasingly judging food imports based on the comparable food safety regulations in their home markets. This is not, by any means, the only option. The Japanese have taken a different approach. Japan (like Singapore) imports most of its food and has designed its food inspection regime to reflect. However, while island states that import most of their food may be able to maintain a firewall-type inspection regime, this is not the North American model.

Like Canada, the United States is an import–export food economy, whose food safety regulator acknowledges the importance of recognizing and using the expertise of nondomestic regulators. Thanks to consumer–industry demand, foreign producers sold over twice as much fresh produce in the United States in 2007 as in 1998. According to the 2010 Georgetown University report, trade agreements—like the WTO—is a prime factor in the increase (Produce Safety Project, 2010).

In a perfectly competitive globalized world without food safety restrictions, agribusiness would be able to import and export products to all countries. Thankfully, food safety regulations are set for the world that exists, if not the ideal one. Still, despite national food safety regulators, agribusiness is expanding international trade. The examples of Singapore and China, Canada and its worldwide imports, and the United States with its fresh produce all point to a shift in the food production map. While each of those examples demonstrates a domestic food safety agency that is using a different strategy to ensure the safety of imported food, none of these countries are becoming more protectionist under the guise of food safety.

The decrease in protectionism reported in the *Food Safety Performance World Ranking Initiative* is a blunt instrument to dissect food safety regulations, international adoption of best practice, and the respect that a country's food safety authority has for its international peers. Admittedly, a high protectionist rating could also reflect a country's concern about the safety of its own, uncompetitive agricultural industry, rather than the safety of its food. Still, as revealed in the *Food Safety Performance World Ranking Initiative* study, protectionism is falling among the 17 countries under study. Moreover, the Singapore–China, Canada, and United States examples all point to outward-looking food safety regulators and expanding agribusiness options built around trade.

Trade, like the other three criteria, deals with a special component of the industry–regulator relationship. Like producer–consumer and regulator–consumer, the interplay between these different stakeholders forms an important part of the holistic industry. The importance of the agribusiness sector and its industry interests should not overshadow the public policy focus and State–Pressure–Response model adopted in the *Food Safety Performance World Ranking Initiative.* The Governance and Recalls evaluation deals strictly with the policies and outcomes adopted by countries and their food safety regulators. Nevertheless, in studying, evaluating, and ranking each country's performance in this metric, intuitive and observed business impacts should also be noted. Whatever importance one places on the interests of producers in the food sector, they clearly play an important role. Moreover, while Governance and Recalls and the *Food Safety Performance World Ranking Initiative* center on food safety, these findings also say something about the international food business.

6

Traceability and Management

The Traceability and Management section of the *Food Safety Performance World Ranking Initiative* is unique in this study as it only comprises one outcome. This focus on traceability underscores the importance on a country's food safety system that a deep and effective traceability system can have. From a Canadian perspective, this is also an area where Canada significantly lags behind its *Food Safety Performance World Ranking Initiative* peers, although it performs no better or worse than the United States.

In addition to exploring the *Food Safety Performance World Ranking Initiative* ranking, criteria, and country-by-country data, this chapter also explores the importance of the international CODEX Alimentarius standard, particularly as it applies to issues of food system tractability. Also covered are the Canadian peculiarities, the role of business in setting up a traceability system (should it be organized from a top-down or bottom-up perspective?), and two examples of Canadian food crisis and the role that our traceability system played. Finally, with the modern, globalized food industry, traceability (like many food safety concerns) is closely related to questions of trade, technology, and the evolution of food safety systems in the future.

Traceability and Management in the *Food Safety Performance World Ranking Initiative*

The investigation into Traceability and Management comprised one outcome: the depth of traceability systems in the food chain. Countries were graded based on their performance in this criterion on a sliding scale from superior to poor based on the comprehensiveness of the individual country's traceability system. Within these grades, the depth of the traceability system was evaluated, with better systems earning their country a higher ranking.

Food Safety, Risk Intelligence and Benchmarking, First Edition. Sylvain Charlebois.

In addition to the comparative depth of a country's traceability system, country-specific information was considered. For example, Japan imports about 60% of its food (Japan Ministry of Health, Labour and Welfare, 2010); therefore, its traceability and management systems *should* be different from those in other countries. Australia has a well-established (albeit limited) traceability system, which earned it a higher ranking. Alternatively, Sweden's system is relatively new, lowering its comparative ranking.

As illustrated in Table 6.1, Austria, Belgium, Denmark, Finland, France, Germany, Ireland, Italy, the Netherlands, Sweden, and the United Kingdom (all European Union (EU) member countries) earned Traceability and Management grades of superior. As a result of the EU rules for domestic traceability systems, each of these countries has a

Table 6.1 Ranking for traceability and management.

			2008 Comparison	
Rank	**Country**	**Grade**	**Rank**	**Grade**
1	Finland	Superior	4	Superior
2	Denmark	Superior	7	Superior
3	Italy	Superior	3	Superior
4	Belgium	Superior	5	Superior
5	United Kingdom	Superior	1	Superior
6	Austria	Superior	8	Superior
7	Australia	Superior	2	Superior
7	Netherlands	Superior	11	Average
9	France	Superior	9	Average
10	Japan	Superior	6	Superior
11	Sweden	Superior	14	Average
12	Germany	Superior	10	Average
13	Ireland	Superior	15	Poor
14	Norway	Average	17	Poor
15	Canada	Poor	13	Average
16	United States	Poor	16	Poor
*	Switzerland	N/A	12	Average

Increased grade	No change	Decreased grade	No data

comprehensive (covering all types of food and food-input products) system of traceability from farm to fork. Australia, Japan, and Norway each scored grades of average for their established but nonuniversal farm-to-fork traceability systems for some food products. Canada and the United States do not have well-established farm-to-fork traceability systems for any food product—although both are working on creating one. Therefore, they earned grades of poor.

Switzerland does not have any data on a traceability system; therefore, it earned a grade of not applicable and no rank.

Comparison with 2008

France, Germany, Ireland, the Netherlands, Norway, and Sweden all improved their grade from 2008. With the exception of Norway, all of these countries are members in the EU, which, by adopting EU-required traceability systems, earned them grades of superior. Norway, because of its EU-inspired traceability system (required for trade through the European Free Trade Association (EFTA)), improved its 2008 grade of poor to a 2010 grade of average.

Canada was the only country to earn a lower grade in 2010 compared with 2008. While it does have tracking systems for its livestock industry, it is still developing a farm-to-fork traceability system. As a result, it shares a grade of poor with the United States (which has a similar, in-progress system under development).

Analysis of Ranking Data

Depth of Traceability Systems in Food Chain

Traceability and Management was measured by the depth of the domestic system for tracing food products through the food chain. The international foundation for traceability standards is set by CODEX. Its *Principles for Traceability/Product Tracing as a Tool within a Food Inspection and Certification System* standard is supposed to "contribute to the protection of consumers against food-borne hazards and deceptive marketing practices" and to facilitate "trade on the basis of accurate product description." In this sense, traceability is closely tied to recall effectiveness and the regulation of food labeling practices. Codex recognizes that "[traceability] does not in itself improve food safety outcomes unless it is combined with appropriate measure and requirements." In fact, it only contributes to the "effectiveness and/or efficiency of associated food safety measures" (CODEX Alimentarius Commission, 2006).

The goal of the CODEX standard is to "follow the movement of a food through specified stage(s) of production, processing and distribution." Specifically

> The traceability/product tracing tool should be able to identify at any specified stage of the food chain (from production to distribution) from where the food came (one step back) and to where the food went (one step forward). (CODEX Alimentarius Commission, 2006)

As a result, the measurement of traceability system depth in this study focused on the level to which each of the countries has achieved comprehensive farm-to-fork systems. Countries with complete traceability systems (regardless of their complete impact on food safety) were given higher grades than those countries with only partial traceability systems (e.g., livestock only). Countries with well-below-average or in-progress systems earned the lowest grades.

The direct comparison was between whether a country has a universal, holistic traceability system for all food products required by law or regulation. The ideal model is a farm-to-fork system. Countries with these universal systems received grades of progressive. Countries that offer farm-to-fork systems for only some products (e.g., livestock) received grades of moderate. Countries with incomplete or in-progress traceability systems received grades of regressive.

The EU countries (Austria, Belgium, Denmark, Finland, France, Germany, Ireland, Italy, the Netherlands, Sweden, and the United Kingdom) all received grades of progressive for the universal EU-regulated farm-to-fork traceability system. Australia, Japan, and Norway each received grades of moderate for regulated farm-to-fork traceability systems for their livestock. Canada and the United States each received grades of regressive for their incomplete or in-progress systems. Switzerland does not have any information of a traceability system.

Comparison with 2008

As illustrated in Table 6.2, Sweden was the only country to raise its grade. The universality of EU traceability requirements ensures that EU member Sweden is part of the progressive (model) traceability regime. Compared with the comprehensive farm-to-fork EU system, Australia, Canada, and Japan received lower grades in 2010 than in 2008. Australia and Japan simply do not provide a comprehensive traceability system, limiting their (mandatory) efforts to livestock. Canada, which similarly limits its traceability system, is only now implementing a fork-to-fork model.

Table 6.2 Depth of traceability systems in food chain—2008 and 2010 grades.

Country	2008 Grade	2010 Grade
Australia	Progressive	Moderate
Austria	Progressive	Progressive
Belgium	Progressive	Progressive
Canada	Moderate	Regressive
Denmark	Progressive	Progressive
Finland	Progressive	Progressive
France	N/A	Progressive
Germany	N/A	Progressive
Ireland	Progressive	Progressive
Italy	Progressive	Progressive
Japan	Progressive	Moderate
Netherlands	N/A	Progressive
Norway	N/A	Moderate
Sweden	Moderate	Progressive
Switzerland	N/A	N/A
United Kingdom	Progressive	Progressive
United States	Regressive	Regressive

Increased grade	No change	Decreased grade	No data

Investigation on Traceability and Management

The investigation on Traceability and Management comprised one outcome: the depth of traceability in food chains in the 17 countries. Countries with comprehensive traceability systems provide food safety regulators with the ability to accurately and quickly track problems and pull unsafe products from the food chain.

In Canada, traceability is enforced by the CFIA (Health Canada, 2010), which maintains animal identification programs for beef, dairy, bison, and sheep. These regulated animals have registered identification tags, allowing the CFIA to trace the origins of "tagged animals involved" in the event of a food safety emergency. However, the administration of traceability investigations is a shared responsibility with the provinces. The two levels of government and industry are "phasing in" a National Agriculture and Food Traceability System, beginning with livestock and poultry (Canadian Food Inspection Agency, 2010b). One of the goals of this project is to develop the capacity to "trace products/attributes along the farm-to-fork continuum." This is an in-progress evolution program, and there are no benchmarks or expected dates of completion.

Like Canada, traceability in the United States is generally an industry responsibility. While the US government has a "long, albeit limited, history" of requiring traceability, "U.S. traceability systems tend to be motivated by economic incentives, not government traceability regulation." As a result, there are broad differences in the breadth, depth, and precision of these systems (Golan et al., 2004). In February 2010, the USDA began an Animal Disease Traceability Framework program, which will regulate interstate livestock. This program will also partner with subnational governments to regulate within these individual jurisdictions (United States Department of Agriculture: Animal and Plant Health Inspection Service, 2010). Nevertheless, compared with its peer countries, the United States (like Canada) does not have sufficient depth to be considered anything but regressive in depth of traceability.

While not on the level with the EU's comprehensive farm-to-fork model, Australia has a hybrid federal, state, and industry traceability system that focuses on livestock. Operating "from paddock to plate," Australian food standards cover the whole supply chain. Set by the Food Standards Australia New Zealand (FSANZ), traceability policies are administered by SafeMeat, a partnership between industry and the federal and state governments (SafeMeat, 2010). According to SafeMeat, "it is a requirement that [live]stock arriving at export abattoirs can be traced to the last property of residence." Traceability for beef and sheep meat through the National Livestock Identification System has been mandatory since 2005 and 2006, respectively (Meat & Livestock Australia, 2006a; 2006b). This system is being expanded to include sheep, goats, and pork (SafeMeat, 2009), which were required to carry tags as of 2008 and 2009. Moreover, Australia has also required a Property Identification Code (animal identification tag) since the 1960s. Finally, the country also has National Traceability Performance Standards, which the Primary Industries Ministerial Council (a federal–state cooperation agency for agriculture has endorsed, to measure the success of Australia's traceability system.

Similar to Australia, Japan also has a mandatory traceability system for cattle and beef, with a voluntary traceability system for other food items run by industry. Before June 2003, Japan historically used the Hazard Analysis and Critical Control Points (HACCP) and International Organization for Standardization 9001 systems. Based on these systems and the CODEX framework, Japan built its new traceability system in the wake of food safety scares, including bovine spongiform encephalopathy

(BSE) ("mad cow" disease). The goal of the Japanese system is to provide a comprehensive food traceability system, covering "the movement of any food product." However, this system has begun with a limited scope, which is expected to broaden. Given the disparity between regulated traceability and voluntary traceability, different food industries are legally required to have varying levels of traceability (Japan Revision Committee on the Handbook for Introduction of Food Traceability Systems, 2007).

Like Japan, Norway has an animal and animal product traceability system, which requires that an owner knows where the animal or product came from and to whom it was sold. However, these rules exist "primarily" because of requirements set by the EU, and less for domestic concerns (Norway Ministry of Agriculture and Food, 2010). The power to require traceability in food products also exists in the Norwegian *Food Act* (Norway, 2005); however, there are currently no regulations to require it (Norway Universitetet I Oslo, 2010). Norway also requires that animals, animal products, and food are registered with the TRAde Control and Expert System (TRACES) database in the EU (Norwegian Food Safety Authority, 2010). (TRACES is part of the EU's internal animal traceability system (European Commission, Directorate-General for Health and Consumer Protection, 2007).) Norway is also a member of the EFTA, which provides access for Norway to the EU's internal market (EFTA Surveillance Authority, 2010b). This participation requires that Norway adhere to EU legislation on traceability (EFTA Surveillance Authority, 2010a).

Australia, Japan, and Norway each have comprehensive and established traceability systems in place (unlike regressive-graded Canada and the United States); however, their systems are not comprehensive—tracing all products from farm to fork.

The EU has specific and comprehensive rules requiring the traceability of food. Specifically

> The traceability of food, feed, food-producing animals, and any other substance intended to be, or expected to be, incorporated into a food or feed shall be established at all stages of production, processing and distribution. (European Parliament and European Council, 2002)

This universal mandatory system of traceability requires that industry be able to identify the source of any food products purchased and the business any food products were sold to and that any products are labeled to "facilitate its traceability." However, the EU does not enforce these rules or punish offenders; these responsibilities are delegated to the individual member states. Generally, the EU does require that these traceability rules are applied to food entering the union; however, this is subject to any unique trade agreements between the EU and the trading nation (European Parliament and European Council, 2002).

It is important to note that the only comprehensive systems are in countries that are members of the EU. Its farm-to-fork model of comprehensive traceability is the standard by which all other countries were measured. Despite the EU's high standard, it is important to recognize that traceability requirements "should not be more trade restrictive than necessary." In fact, they "should be practical, technically feasible and economically viable" (CODEX Alimentarius Commission, 2006). In other words, at a certain point, more traceability depth (information) could become redundant or unnecessary. Moreover, such requirements should not be used as trade barriers. A potential example of the trade requirement impacts that a traceability system can have is Norway (a non-EU member), which has adopted the legal power to enact livestock and food traceability systems and admits that part of its motivation is to continue to trade through the EFTA deal with the EU (Norway Ministry of Agriculture and Food, 2010).

In effect, traceability systems do not have to be identical to be effective: While the EU countries all score the same (by virtue of identical rules), Norway's different system could be equally as good. So, while countries earned the same traceability depth grade, they are not (necessarily) equivalent. The overall Traceability and Management ranking reflects these differences.

Discussion

In order to rank each country's performance fairly and transparently, clear and objective standards were necessary. While the standards in each section of the *Food Safety Performance World Ranking Initiative* are different, establishing or identifying a global best practice aides both the fairness and transparency of the grading system in addition to providing a solid basis for future evaluation or benchmark for the current analysis. In the case of Traceability and Management, the CODEX standard—Principles for Traceability/Product Tracing as a Tool within a Food Inspection and Certification System (CAC/GL 60–2006)—provided this foundation.

According to this FAO–WHO partnership, "The CODEX Alimentarius, or the food code, has become the global reference point for consumers, food producers and processors, national food control agencies and the international food trade." Created in 1961, the CODEX standards have become both the repository of global best practice for food safety—as well as the measuring stick that accompanies that role—and the venue for discussion and debate of the evolving standards that these modern scientific regulations require (World Health Organization (WHO) and Food and Agriculture Organization of the United Nations (FAO), 2006).

The existence of food safety regulations extends at least as far into history as the Assyrians (over 3,000 years ago), with other ancient examples coming from the Egyptian, Grecian, and Roman civilizations. While the long-standing human practice to create local food safety customs and standards (e.g., the 1516 Bavarian Purity Law for beer) were surely important to local producers and consumers, the need for broad scientific regulations on an international scale was a development rooted in the progress of transportation and trade (WHO and FAO, 2006).

The nineteenth century saw numerous technological and scientific breakthroughs, including the introduction of modern regulations and regulatory agencies, as well as the beginning of real international trade in foodstuffs. Along with the scientific recognition of food chemistry as a serious field of study, the development of scientific purity standards and the introduction of canning, an overhaul of the government's role in regulating food safety joined this forward march of producer power. Finally, as a result of transportation improvements and the subsequent increase in international trade, the import and export of food began to expand what were once local issues of food safety into the international scene. For example, by the mid-1800s Europe was importing bananas from equatorial producers, and, only a few decades later, frozen meat made the colonial journey from Australian and New Zealand to the United Kingdom (WHO and FAO, 2006).

While the increased potential of trade from the nineteenth century onward provides a boon to producers and consumers, government food safety regulators realized the dangers that unsafe food imports could have on their citizens (and, presumably, the dangers that unsafe food exports could have on their producers' foreign markets). Therefore, the creation of a body of unified standards nationally (through increased food safety laws, regulations, and agencies) and internationally (a CODEX-like organization) seems not only logical but also inevitable.

The modern FAO–WHO CODEX Alimentarius takes its earliest inspiration from an Austro-Hungarian Empire standard from the turn of the twentieth century called the "CODEX Alimentarius Austriacus." In addition to sharing a name with its Central Europe predecessor, the modern CODEX standard also lacks direct legal force. As a creation of the WHO and the United Nation's FAO, CODEX Alimentarius is a moral rather than judicial force, and its CODEX standards are guidelines, rather than rules that sovereign states must adopt. However, the international makeup of the CODEX Commission (rule-making body), strong scientific focus, and widespread recognition give the standards probative value. Furthermore, these standards are recognized by other international organizations and in international law; for example, "CODEX standards have become the benchmarks against which national food measures and regulations are evaluated within the legal parameters of the World Trade Organization Agreements" (WHO and FAO, 2006).

The recognition of CODEX standards as the benchmark of international best practice in food safety is a by-product of the importance and relevance of its standards, the respect that the organization commands among government and nongovernment organizations, and the need for international standards in a globalized food network. As a result, the *Food Safety Performance World Ranking Initiative* takes account of the importance of CODEX standards as they concern traceability. However, as the results demonstrate, some countries are falling below the ideal farm-to-fork standard.

Canada's Traceability Unpacked

As noted during the *Food Safety Performance World Ranking Initiative* discussion, the CODEX standard for Traceability and Management is to "follow the movement of a food through specified stage(s) of production, processing and distribution." Specifically, the traceability program should "identify at any specified stage of the food chain (from production to distribution) from where the food came (one step back) and to where the food went (one step forward)" (CODEX Alimentarius Commission, 2006). Canada performed poorly on the CODEX-inspired standard of having a farm-to-fork traceability system. In fact, the national CFIA does not have sole jurisdiction of the existing beef, dairy, bison, and sheep animal identification program (Canadian Food Inspection Agency, 2010b), nor have benchmarks or deadlines been set for its new farm-to-fork National Agriculture and Food Traceability System.

Nevertheless, the Canadian reality is perhaps not as grim as the *Food Safety Performance World Ranking Initiative* would suggest. The initiative only looked at national-level food safety organizations and projects, in part to ensure fair comparisons between each country's national food safety organs. As a result, Canadian traceability projects undertaken at the provincial level did not affect the national score. Fortunately, provincial projects and shared federal/provincial programs are moving Canada toward the CODEX traceability ideal.

In July 2008, the federal, provincial, and territorial agriculture ministers settled a framework agreement for a $1.3 billion (Canadian) Growing Forward program. As part of this shared-cost program (a 60/40 split between the national and subnational governments), the signatories promised to "[focus] on supporting the enhancement of food safety systems and the continued implementation of on-farm environmental actions" (Agriculture and Agri-Food Canada, 2008). In addition to reflecting the Canada-specific divisions of powers and labor between national and subnational governments, this agreement also highlights the important role that producers play in traceability specifically and food safety generally. (This topic is discussed further later.)

As part of the federal contribution to the Growing Forward program, the Canadian Integrated Food Safety Initiative covers food safety through two components:

- The Food Safety Systems Development (on-farm and postfarm producer HACCP systems)
- The Canadian Industry Traceability Infrastructure (which includes the National Agriculture and Food Traceability System)

The latter project, the Canadian Industry Traceability and Infrastructure program, directs money toward industry organizations of up to $2 million (Canadian) covering up to between 50 and 90% of traceability project costs.

At the provincial level, each government has set up its traceability-centered programs differently (presumably and hopefully to suit the individual needs of the 10 different agricultural sectors). While each province is participating in the Growing Forward program, the province-by-province setups vary. For example, in Prince Edward Island (PEI) the provincial government will "provide assistance to commodity boards/associations to enable their industry to participate in a full traceability system" (Prince Edward Island Department of Agriculture, 2010). However, rather than following the intermediary approach found in PEI, the Government of Ontario has divided its food safety and traceability programs between business funding (cost offsets) and "sector-level education and outreach activities." While many of these provincial programs are relatively open ended, in as far as they deal with the particular on- or off-farm traceability system, for its part, the Saskatchewan Ministry of Agriculture has focused its traceability program specifically on the introduction, training, and use of radio-frequency identification (RFID) tags and associated hardware. Finally, not to be outdone, Alberta Agriculture and Rural Development has designed its program to serve all of the stakeholders that its provincial counterparts have focused on individually, offering programs in RFID implementation, nonspecific traceability projects, and general training programs (Alberta Agriculture and Rural Development, 2010).

While each province presumably services its agricultural sector to suit its unique needs, the Alberta and Saskatchewan support for RFIDs comes at a valuable time for the cattle industry as the CFIA has switched to a mandatory RFID tracking system for livestock. As of July 1, 2010, all cattle must be RFID tagged before moving farms or lots (Canadian Food Inspection Agency, 2010a). The switch is the culmination of a four-plus-year phase-in period as Canada has been moving away from bar-coded tags, which have been unavailable for purchase since 2006.

In addition to the variety of traceability programs offered at the federal and provincial levels through Growing Forward, some provinces have taken it upon themselves to create stand-alone traceability projects. One example is from the Ontario Ministry of Agriculture, Food and Rural Affairs, which has created a series of Advantage food safety programs to support a CODEX-like system of Ontario-specific benchmarks based on HACCP and other standards. In addition to encouraging traceability and food safety standards in Ontario, producers can become "Advantage certified" by the Ministry.

While the creation of a farm-to-fork traceability system in Ontario may reflect progress toward solving the Canadian deficiency in Traceability and Management noted in the *Food Safety Performance World Ranking Initiative*, it is voluntary progress. Moreover, so too are all of the Growing Forward initiatives. Setting aside whether the government—either at the federal or provincial level—should require a full-borne farm-to-fork traceability system, the current model places enormous responsibility on the shoulders of Canadian producers. To wit, while Canadian governments are willing to foot much of the bill, currently, Canadian producers must voluntarily create and adopt traceability programs.

The Role of Business: Top-Down or Bottom-Up Traceability

As part of its Growing Forward agreement, the governments of Canada committed to "continue to work with the sector to put in place traceability and biosecurity systems" (Agriculture and Agri-Food Canada, 2008). While "work with" is somewhat vague, the voluntary-based traceability offers discussed earlier suggest that these governments see traceability and the creation of a Canadian standard as something of a bottom-up process.

This may pose something of a problem for quick improvement in Canada's Traceability and Management grade in future *Food Safety Performance World Ranking Initiative* projects. While the initiative's focus on national food safety agencies might undercount some Canadian progress made through its system of federalism, this collaborative approach between different levels of government is in direct contrast to the EU's approach: legislate and require the member states to administer comprehensive farm-to-fork traceability (European Parliament and European Council, 2002).

The EU's success in creating a comprehensive farm-to-fork traceability system reflects one of the strengths of a top-down model. Similarly, most of the successful standards rely in some part on a directive. Even the international CODEX Alimentarius standards are the result of explicit mandates given by the WHO and FAO (2006).

Whether the industry (even through producer, sector, or industry associations) can create effective and comprehensive farm-to-fork traceability standards in Canada, logistical and technological issues will become important first steps. Nevertheless, the sheer complexity of the undertaking is a worthwhile public policy question for Canadian agricultural policymakers. Consider that food safety issues can arise at any stage in the variety of steps between the farm and fork. In addition to the four basic food-stage handlers (farmers, producers, retailers, and consumers), modern trade and production patterns mean that some or all of these stages may take place in other jurisdictions, food safety systems, or countries (Buckley and Reid, 2010). The complexity of farm-to-fork traceability comes into even sharper relief when one considers the logistical difficulties Canada has had in establishing and administering (comparatively) simple one-dimensional traceability systems. The recent Canadian crises of the 2008 listeriosis outbreak and the 2003 BSE cow offer important examples of how far Canada has to go in building farm-to-fork traceability.

2008 Listeriosis Outbreak

The 2008 listeriosis outbreak involving Maple Leaf Foods products is an excellent example of the importance of traceability to producers, consumers, and governments. One of the (57) recommendations in the *Report of the Independent Investigator into the 2008 Listeriosis Outbreak* suggests that "Federal, provincial and territorial governments and their research funding agencies should initiate and support further research into ... improved traceability technology and methodology" (Weatherill, 2009). Doubtlessly, an outbreak that claimed the lives of 22 Canadians and represents the biggest food safety story Canada has seen (Goveia, 2010) is an excellent opportunity for a food safety system overhaul. However, the traceability-specific details of the crisis are somewhat more nuanced.

In early July 2008, the Public Health Agency of Canada received two listeria specimens from Ontario patients. A few days later, two more Ontario patients also became ill with listeriosis. While Toronto Public

Health began an investigation into the most recent two cases, no agency made the connection between those cases and the two reported a few days earlier. Soon, the Public Health Agency of Canada recognized that the original two listeria specimens were DNA matches—meaning that those cases were linked. While Toronto Public Health and the Public Health Agency of Canada continued their two separate investigations, it took until July 25—2 weeks after the first recoded cases—that the Ontario Ministry of Health and Long-Term Care detected an increase in the number of reported listeriosis cases. Nevertheless, it was not until August 7 that the CFIA linked the various listeriosis cases to Maple Leaf Foods-produced deli meats and initiated a food safety investigation. By August 12, the Public Health Agency of Canada had identified DNA links between listeria specimens from Ontario, Quebec, and Newfoundland and Labrador. On the same day, the CFIA tested a Sure Slice-branded Maple Leaf Foods product that would be found to carry *Listeria*. Once the test results arrived on August 16, the CFIA issued a Health Hazard Alert for two Maple Leaf Foods products. By August 23, Maple Leaf Foods was recalling all 191 products from its Barton Road plant in Ontario where the outbreak originated (Weatherill, 2009).

At first blush, traceability is not the cause of the listeriosis crisis, nor a principle problem. In fact, traceability was not one of the four critical weaknesses noted by the independent investigator in the Maple Leaf Foods case. The problem, for almost two months between early July and the end of August, was neither tracing products nor instituting a recall. Instead, the involved food safety authorities were trying to find out what was making people sick. However, one of the problems with the prerecall investigation was the absence of complete and centralized data (Weatherill, 2009). This problem, at its basic level, does represent a traceability concern: When basic information is not available, "the original sources of small outbreaks often go unidentified, allowing contaminated food to continue to enter the production system and small problems to grow more serious" (Buckley and Reid, 2010). Furthermore, as Maple Leaf Foods products were recalled, the CFIA and local public health authorities undertook over 29 000 verification checks to ensure that the 24 products (two Maple Leaf Foods products and 22 items made from affected Maple Leaf Foods products) were pulled from shelves. These recall checks "put a tremendous strain on all involved," and "while these activities were necessary, there was very little information" (Weatherill, 2009).

In effect, during the prerecall and recall stages of the listeriosis crisis, the Canadian food safety system provided insufficient information to (first) comprise a traceability system and to (then) comprise an effective traceability system. The absence of basic information certainly made the prerecall crisis more confusing, reflecting an incomplete or ineffective traceability system. Later, during the recall stage of the crisis, the needs of the deli-meat traceability system overwhelmed the regulators' ability to do their jobs. Under the traceability system as it applied in this crisis, the regulators had to make 29 000 checks to ensure that retailers received notice of the producer product recalls and actually removed the affected products. Depending on how one looks at this situation, it either represents one step (producer to retailer) or two steps (producer to retailer and retailer to consumer) on the four-party farmer–producer–retailer–consumer continuum. Either way, the fact that Canada's food safety s by the necessary requirements of the traceability system (to verify that the recalls were proceeding) is a black mark on Canada's packaged-food traceability capability.

2003 BSE Cow

Like the listeriosis crisis five years later, the May 2003 announcement that a 6-year-old Canadian-born Angus cow had tested positive for BSE had a profound impact on the country's food safety confidence, particularly on the beef industry and its trade opportunities. Shocks like this can result in a founding act (Pauchant and Mitroff, 1992), which can trigger a crisis that results in a total breakdown of the collective sensemaking of a marketing channel (Pearson and Clair, 1998). However, unlike the 2008 listeriosis crisis, the country's beef traceability system performed remarkably well.

This was not the first BSE-positive cow found on Canadian soil; in 1993 a British-born cow was found on a Red Deer, Alta., farm. In response, Agriculture Canada destroyed the animal and five of its herd mates. However, the Canadian-born cow 10 years later caused a massive disruption to the Canadian beef market both domestically and internationally. This time, the BSE case caused disruptions to Canadian beef and cattle exports destined for around the world, including the United States, Japan, Mexico, and Thailand. With a markedly different sense of urgency than shown in 1993, the CFIA immediately started an investigation while destroying and testing 1400 cows.

There are a few issues to discuss concerning the disparity between the 1993 and 2003 BSE cases. First, the 2003 cow was to be slaughtered at a provincially licensed slaughterhouse. Second, meat from provincially licensed slaughterhouses is only saleable inside the licensing province. Therefore, third, meat from the BSE-infected cow was not destined for international export. However, the international concern—despite assurances to the contrary from government officials—was that the Canadian beef sector had other BSE-infected animals working its way through the system that could be destined for export. (As it turned out, the fear of a Canadian-born BSE-infected cow being exported should have been just as much a concern as, in retrospect, importing beef from Britain in 1993: In December 2003, a Canadian-born cow tested positive for BSE in Washington state.)

In the immediate aftermath, the damage to the Canadian industry and individual producers was enormous. In 2002, the Canadian beef industry was worth over $7 billion (Canadian), with the country running a $3 billion (Canadian) trade surplus in beef products. After the May 2003 BSE announcement, a cow that would have normally sold for $1300 (Canadian) would only fetch $15 (Canadian); the domestic combination of insufficient demand (for export-intended meat) and concerns about the security of the industry resulted in a glut of oversupply. In desperation, Canadian beef producers wanted to slaughter 620 000 cows (of about 13.5 million) to restore the industry's economics. Instead, the federal government paid BSE-related aid to the industry of $190 million (Canadian) in 2003, over $1.1 billion (Canadian) in 2004, and $321 million (Canadian) in 2005. Still, despite all of the damage to the industry and to Canada's reputation as a source of safe agricultural products, it is important to recognize that the country's traceability and management systems performed quite well.

In fact, the Canadian food safety system received high praise. A June 2003 report by a team of foreign experts—"Report on Actions Taken By Canada in Response to the Confirmation of an Indigenous Case of BSE"—begins by highlighting:

> The openness, full disclosure and access to personnel provided by the Canadian authorities to our team, the international community and the public. The approach to sharing of information and communication demonstrated by Canada is a model to be emulated. (Kihm et al., 2003)

In the weeks since the May 2003 announcement, the team was "impressed" with the "comprehensive scope, level of analysis and thoroughness of the investigation to date." Furthermore, Canada had collected and analyzed an amount of data that exceeded investigations in other BSE-affected countries. While the final report highlighted a number of issues, their policy recommendations for traceability encouraged the "continuous investment, improvement and extension of the cattle identification system," which—had it provided data extending 7 years—would have been helpful in this case. However, by 2003, the CFIA was already 3 years into its traceability implementation program. While the expert report found that Canada's traceability regulations in 1997 (7 years before the incident) were insufficient to track the BSE-infected cow's history, by 2003 the CFIA had implemented sufficient programs (Kihm et al., 2003). Like the 2008 listeriosis crisis, the postcrisis report recommended improvements to the traceability system, but no major problems.

Nevertheless, the absence of adequate traceability information left the CFIA's final report on the BSE case with some ambiguity. To wit, the CFIA was only 95% sure that the BSE-infected cow was born in Canada and could only establish that it "was between six and eight years old" (Canadian Food Inspection Agency, 2003). In reaction to the expert report and the CFIA's findings, the Canadian Standing Committee on Agriculture and Agri-Food recommended that the Canadian Cattle Identification Program be "enhanced" with a "comprehensive" traceability system. Its reasons are as follows:

> On 20 May 2003, when news of Canada's BSE case became public, the CFIA undertook immediately to trace the cause of the BSE to the source farm. While this was done with considerable speed, the Standing Committee believes that the use of new technologies could have substantially reduced the time required to complete this process. The Standing Committee believes, therefore, that special attention should be given to implementing a program that would facilitate the immediate trace-back of animal history and lineage in instances of serious communicable diseases. In doing this, all efforts should be made to ensure that any proposed national system incorporates existing regional and national practices along with new technologies and animal health procedures. Furthermore, when devising such a national traceability system, the government should be mindful that all Canadians would benefit from effective tracking of animals, and program costs should be distributed accordingly. Such a system would provide economic and health benefits, and be a mechanism to promote and demonstrate superior food security both to Canadians and to our trading partners abroad. (Canada Standing Committee on Agriculture and Agri-Food, 2003)

In other words, while no agency or expert panel wanted to blame the beef traceability system, everyone recognized the importance of maintaining and upgrading the system. Nevertheless, the concerns raised by the 2008 listeriosis crisis and the 2003 BSE cow represent legitimate points of concern for the existing top-down traceability systems in Canada.

Traceability and Trade

While the cost of establishing a Canadian farm-to-fork traceability system may seem high, the importance of traceability is quickly growing. Consider that the birth of the modern international CODEX system of practices is an outgrowth of scientific capability (understanding food safety issues) and technological progress (new methods of production and trade). Recalling the big changes of the nineteenth century, transportation and production innovations allowed for canned foodstuff and the import of fresh nonnative fruit and frozen meat products across oceans (WHO and FAO, 2006). In the 100 or so years since these transoceanic journeys, nothing changed has changed: Producers still want new markets, consumers still want variety, and technical, scientific, and social changes are making international food trade easier.

The process of globalization is dramatically affecting even mature agricultural economies. Since the 1990s, the United States has imported substantially more fresh produce than it used to. While this might seem strange considering that amount of fresh product exports from the United States that Canadians habitually find in the grocery store—Georgia peaches, Idaho potatoes, and California and Florida citrus, the increase in US imports is largely a result of increased variety. For example, it now imports about 100% of its bananas. While international trade in this particular fruit has played a significant role in the relationship between the United States and its Latin American neighbors, the banana is now imported in significant enough numbers to be sold year-round and popular enough to remain the best-selling fruit in the country. Furthermore, while the import of kiwis (55%), pineapples (75%), or plantains (100%) is no surprise, the share of imported fruits and vegetables that could be grown in the United States includes bell peppers (31%), blueberries (33%), cantaloupes (30%), cucumbers (48%), table grapes (40%), tomatoes (35%), raspberries (36%), and tangerines (25%). Perhaps most shockingly, the United States also imports about 49% of its onions (Produce Marketing Association, 2007). This last item is particularly interesting because the popular Vidalia sweet onion is an example of a region-specific (Vidalia, Ga.) product in the United States that Canadians know well as an import of their own. While the trade in fruit and vegetable imports is quickly growing (from $5.9 billion (US) in 1998 to $13 billion (US) in 2007), the root causes include consumer demand for a greater variety of products, increased globalization causing a reduction in the seasonality of products (many of which are now available 12 months of the year), and improved shipping and transportation options (Produce Safety Project, 2010).

So, perhaps nothing has changed since the 1800s except the pace. While nineteenth-century European consumers wanted bananas and Canada's Commonwealth partners exported frozen meat to the United Kingdom, food trade in the twenty-first century is an important and growing business. However, along with the benefits to consumers and industry, the increase in the global food trade brings many and increasing dangers:

> Historically, geographical separation has been a key barrier to the spread of disease, but international trade and travel are reducing this barrier, allowing diseases to move fluidly between regions and continents. Food-borne disease is no exception; 50 years ago, it was prohibitively expensive to transport produce and other perishable foods over great distances. Today these foods can be transported halfway around the world, allowing consumers to enjoy a variety of fresh products year-round. While rapid enough to bring foods to consumers before spoiling, long-distance shipping tends to increase the amount of handling and the time between farm and consumer, which can give pathogens on the food more time to multiply, potentially amplifying the hazard for the consumer. (Buckley and Reid, 2010)

While the *Food Safety Performance Word Ranking Initiative* covers a wide variety of policy solutions and outcomes that the 17 countries in the study have deployed to solve these modern food safety issues, traceability, like food recalls, is a tool deployed after a problem arises. As noted earlier, the US FDA's explanation for why food recalls and food-related illnesses occur is that "Unpredictable events, mechanical and human error, and environmental conditions all play a role in the problems we continue to see in food production, processing, and distribution." Since accidents are unlikely to disappear, effective traceability systems (like food recall regulations) are an important tool, especially in the trade-heavy, globalizing world food industry. Fortunately, the scientific and technological industries that have created or highlighted these modern food safety concerns also provide some help in creating better safety systems.

Technology and the Future

According to the CODEX standard, traceability is an outcome-based undertaking: Can one trace a product backward and forward through the farm-to-fork continuum? (CODEX Alimentarius Commission, 2006)

While this outcome may represent a daunting task, the basic record-keeping idea behind modern traceability standards is older than the Roman Empire (OnTrace Agri-Food Traceability, 2007). Consider that the branding, tagging, and documenting of livestock animals must be nearly as old as their agricultural domestication. If only to establish and maintain proof of ownership, the tracking of animals by producers is an ancient practice.

While the modern traceability standards are beyond branding cattle, technological advances are easing the transition away from the status quo toward real, digital farm-to-fork traceability. For example, in Canada, the CFIA maintains traceability in Canada for beef, dairy, bison, and sheep based on identification tags. The Canadian Cattle Identification Agency (CCIA) (nine provinces, less Quebec) and Agri-Traçabilité Québec (Quebec) hold data regarding specific animals, including premises (location) identification, movement, and slaughter (Canadian Food Inspection Agency, 2010b).

The importance of traceability is clear, especially considering the recent food safety events in Canada. To this end, in addition to improving the amount of data available for livestock (and other food products), technological developments are providing opportunities to make the traceability process easier to implement, less costly, and more accurate. One example of the recent trends in Canada are the several RFID support programs offered by provincial governments as part of their Growing Forward programs as well as the CFIA and CCIA shift to RFID tags over bar codes.

The Alberta Ministry of Agriculture and Rural Development has specifically targeted RFID technology for producers to "increase [their] profitability and strengthen [their] competitive position." Through the increased use of these transmitters, the Ministry hopes to increase livestock movement recording and increase overall traceability adoption in the industry. Targeting "feedlots/backgrounders feeding more than 1000 head of cattle annually," the RFID Technology Assistance Program will cost-share up to $3000 (Canadian) per handheld reader and software application or up to $20 000 (Canadian) per panel reader and software application (Alberta Agriculture and Rural Development, 2010).

In Saskatchewan, the Ministry of Agriculture has a $5-million (Canadian) Voluntary Livestock Traceability Rebate program that provides up to 70% rebates on RFID handheld and panel readers, in addition to support for training, software, installation, and facility modifications relating to their use. This program can be worth up to $50 000 (Canadian) for producers and closely related industry or up to $100 000 (Canadian) for livestock auctioneers. The Government's stated intention is to support the Canadian Cattle Identification Agency's presence in Saskatchewan.

These examples of provincial programs incorporating RFID technology to improve existing databases, support producers who implement improved traceability systems, and facilitate this data collection with greater ease and lower cost reflect the promise that technology presents for the food safety system. In response to the RFID shift by the CFIA and CCIA, Canadian Agriculture Minister Gerry Ritz claimed: "A strong traceability system will help Canadian producers get the premium prices their top quality products deserve around the world. With RFID technology, we'll be better able to trace an animal, which is not only important to human and animal welfare but also key to the sustainability of the Canadian livestock industry as a whole" (Canadian Food Inspection Agency, 2010a).

While livestock traceability is the most developed in Canada, there is no reason why other traceability systems for nonanimal food products cannot be similarly improved. For example, the automated bar-code-based logistical systems of modern courier services effectively provide the same service: tracking a package from Point of Origin A to Destination B. If Canada Post's Purolator service can log the steps necessary to move a book ordered from a Vancouver store through the national postal grid to a purchaser's home, surely this technology can track shipments of PEI potatoes or Quebecois cheese from farm to fork.

One example of the construction and implementation of a broader traceability system is the Ontario Agri-Food Premises Registry. This system gives producers the ability to obtain a unique identifying number "assigned to a parcel of land that is associated with agri-food activities." Funded by the Ontario Ministry of Agriculture, Food and Rural Affairs, the registry is managed by OnTrace Agri-Food Traceability, a "not-for-profit, industry-led organization." OnTrace, building on the Ontario Agri-Food Premises Registry and the Canadian National Agriculture and Food Traceability System project, hopes to create a broader system. In a March 2010 technology vision, OnTrace noted several issues surrounding the increased use of technological solutions to Ontario (and, broadly, Canadian) traceability systems. This included interoperability with different jurisdictions, participation among industry stakeholders at all stages of the food production process, establishment of technology and software standards, security, and cost. (Based on a 2009 USDA report, OnTrace hypothesized that the cost of creating a "full traceability" Canadian database for livestock would be 10% of the $228 million (US) estimate for the United States, that is, $22.8 million (US).) (Albu and Sterling, 2010).

The technological possibilities for traceability specifically and food safety generally are limitless. On the traceability front, increased digitization of records will result in easier accessibility, accuracy, and transparency.

Moreover, with the establishment of national and international standards, data can be shared across jurisdictional lines, aiding in trade and global food security. However, as a first step, traceability and technological standards must be created, adopted by industry, and supported by governments. In Canada, industry and government stakeholders need to decide on how the traceability system should work (top down or bottom up), who should run it (government or delegated industry), and how the division between levels of government will influence the effectiveness of national food safety standards.

EU regulations covering the traceability of all food and animal products of both domestic and imported origin have established the countries adopting EU legislations as strong leaders in global traceability.

Even though Japan's beef labeling law for farm to fork is applicable to domestic products now, the Japanese government has new regulations on rice traceability and various proposed traceability regulations in development for other commodities. This places Japan in a "fast-track" position in food traceability race.

Canada is strengthening its traceability requirements through mandatory livestock identification and swine identification and movement tracking program. The United States is still trailing other nations on food traceability, and the new Food Safety Modernization Act (FSMA) is expected to improve its traceability system.

Australia has a strong cattle and livestock identification system but still needs to develop more advance traceability systems for other domestic as well as imported products. Currently, although many countries lack the specific legislation on traceability, the tracing and tracking of globally imported products are being achieved through record keeping, lot identification, labeling laws, and requirements for importing countries to meet the standards of exporting countries. Reliance on internationally recognized organizations such as GS1 for uniform product coding and on GFSI schemes for the verification of traceability systems has improved confidence in trading partners across different nations.

Even though the traceability of livestock and meat products has improved through the harmonized adoption of animal identification systems, systems allowing for the same level of traceability for other food commodities are still far from being fully implemented. It would be strongly beneficial for global markets to move toward the development of a harmonized and uniform global traceability system by following the examples of EU legislation for the traceability of all commodities and developing electronic identification, database programs, and systems coupled with GS1 identification standards and GFSI auditing schemes. More will be discussed in Chapters 7 and 8.

7

The 2014 Survey

A New Approach

From the 2008 and 2010 reports, we learn a great deal from how countries differ from one another when mitigating risks. We also learn how they value benchmarking reports on food safety performance. Initially, the concept was deemed controversial by many. Since 2008, it seems countries have gotten more comfortable with these reports. But over time, our methodology has evolved which led to a much more robust survey in 2014. Most importantly, the two first reports compelled regulators to organize an international workshop in Helsinki in 2011, in which 21 countries were involved. The workshop allowed the proper evaluation of all metrics used in the first two reports and inspired the generation of a new list of metrics for the most recent survey.

Nations differ not only by their topography, public health challenges, and food supply chain capacity but also by their food safety systems, regulations, and risk intelligence strategies. However different, each country's respective food safety system faces comparable responsibilities to ensure safe food. There is also a real need to harmonize standards and protocols among nations. As a result, routine benchmarking evaluation of food safety performances among nations is essential and appropriate as these nations aspire to improve their food safety system preparedness, responsiveness, and accountability.

There is currently no metric that captures the entire food safety system, and performance data are not collected strategically on a global scale. Increasingly though, food safety regulators from around the world want to know if current domestic policies and regulations are adequate to support food safety risk intelligence efforts. Such efforts are hindered by existing national data, which tend to be segmented, or suffer from limitations such as inaccurate or underestimated data,

Food Safety, Risk Intelligence and Benchmarking, First Edition. Sylvain Charlebois.

information gaps, unfeasible comparisons, absent baselines, or limited access to proprietary data.

Consequently, segmentation and limitations hamper the world's ability to select, build up, monitor, and evaluate food safety performance. As such, benchmarking efforts such as this one can help monitor ongoing food safety performance and inform continued food safety system design, adoption, and implementation toward gradually more efficient and effective food safety policies and practices.

Regrettably, global comparative food safety ranking studies are rare. There are recent performance rankings of specific food safety dimensions, including food traceability (Charlebois et al., 2014b) and food labeling and allergenic risks (Balazic et al., 2013). The Economist Intelligence Unit (EIU) also includes a food safety indicator in its annual Global Food Security Index, but it is limited to three indicators. They are the presence of an agency to ensure the safety and health of food, the percentage of the population with access to potable water, and the presence of a formal grocery sector. In the EIU's view, the latter two indicators assess whether countries have reached levels sufficient to provide safe food. However, none of the three indicators explain or relate to food safety performance.

In sum, there is little published comparable evidence on a broad set of international food safety performance measurements.

Purpose

Previous attempts to rank such broad international food safety performance include two world ranking studies, one conducted in 2008 (Charlebois and Yost, 2008) and the other in 2010. This report builds upon these earlier benchmarking assessments. It makes the case that the use of performance metrics does evolve over as the food safety landscape constantly changes. For example, unlike previous surveys, GFSI has influenced how regulators and industry deal with food safety risks which makes trade pacts less influential, as portrayed in previous surveys.

The report is a comparative study that measures and ranks the food safety performance of 17 peer Organization for Economic Cooperation and Development (OECD) countries: Australia, Austria, Belgium, Denmark, Finland, France, Germany, Ireland, Italy, Japan, the Netherlands, Norway, Sweden, Switzerland, the United Kingdom, and the United States of America.

Benchmarking food safety metrics require quantitative criteria for food safety performance to be set and tracked. However, benchmark assessments or food safety performance comparisons require a robust evidence-based set of valid, ideally objective indicators that can encourage stakeholders to act. Results also need to be understood by the consumer and the broader public to help them appreciate the complexities of the food safety system and how food safety risks are managed for foods and commodities sourced domestically and elsewhere.

Since the initial food safety benchmarking report was published in 2008, the sharing of data and protocols among nations has dramatically increased. The main purpose of this benchmarking assessment is to identify and evaluate common elements among global food safety systems. More specifically, this report identifies those countries that employ comparatively best practices to assess, manage, and communicate the risks related to the safety of food and their respective food systems. The overarching intent of this benchmarking assessment, however, is to stimulate exchange and discussion on food safety performance among nations.

Methodology

The rankings discussed in this report are designed to express the relative strength of each nation's food safety system according to an output benchmarking scale of superior, average, or poor or a response scale of progressive, moderate, or regressive. Output refers to the condition of the food safety performance at the time of the collected data. Response refers to the policies and actions that the country has initiated or will initiate to address food safety. The rankings highlight success and areas for improvement and provide an opportunity to identify national food safety systems that employ effective food safety policies and practices around three food safety risk governance domains: food safety risk assessment, risk management, and risk communication.

The choice of metrics is not trivial when preparing such reports. The data selected for this project were carefully scrutinized. Indicators were selected for their credibility to stakeholders, their capacity to spur stakeholders to action, and thus their ability to leverage the food safety system. Useful indicators must also provide valuable information concerning the performance or status of the particular food safety risk domain (assessment, management, communication), be based upon robust data (i.e., reliable, easily accessible, and readily available), and be sufficiently consistent to allow benchmarking over time and allow for international comparative analysis. Readily available data captured for this study came from publicly available sources from recognized competent authorities in food safety, both national and supranational, such as the Canadian Food Inspection Agency and the European Food Safety Authority.

As with the earlier world ranking studies, this study focuses on indicators that can also be influenced by public policy. In addition, the choice of indicators to evaluate food safety performance has evolved from previous reports, thanks in part to input and consensus from a group of food safety regulators following a food safety metrics workshop held in Helsinki in 2011 (Charlebois and Hielm, 2014), the current availability of data, and efforts to reduce and minimize subjectivity in favor of objective and robust metrics. In all, a total of 10 indicators were selected. They include, by food safety risk domain,

- Risk assessment, a science-based process that assesses exposure and characterizes food safety risks. Includes pesticide use (chemical risk in agricultural production), total diet studies (TDS) (reporting of chemical food hazards), food-borne illness rates (microbial risk), and national food/dietary consumption studies
- Risk management, a policy-based and commercially based process to prevent, control, and mitigate risks while ensuring health protection and fair trade practices. Includes national food safety response capacities, food recalls, food traceability, and radionuclide standards
- Risk communication, the exchange of information and opinions around food safety risks (actual or perceived). Includes food allergies and labeling (allergenic risk) and public trust

Additional food safety performance indicators were considered but discarded owing to their subjective nature; inadequacy to describe performance; or lack of data such as the rate of inspections and audits, food safety education and training programs, and government food safety information of quality and distribution to industry and consumers.

This report includes both output and response indicators that measure results (not efforts) and other indicators that compare national standards. The indicators are based on secondary sources that are available in English and/or French, easily accessible, and readily obtainable through keyword searches on the Internet. Of note, secondary data were not always available for each country studied and time periods differed on occasion.

Food Safety Risk Assessment

Assessing food safety risks is the first step in effective risk governance. One potential food safety performance metric is the number of recorded international food safety violations of traded raw and minimally processed foods.

According to a recent study of 3400 food safety violations that occurred in 2013, the countries with the highest number of violations were India, China, Mexico, France, and the United States (Food Sentry, 2014). No other OECD country appeared in the top 10. The food safety violation results were not normalized on volume (for weighted comparison) or assessed for severity and reflected only the share of foods that were inspected (a small share of the overall food trade). Results revealed, however, that chemical risks (pesticides) accounted for over a third of all violations. This was followed by microbial risks (pathogen contamination at 22%), which, when combined with food safety chemical risks, represented more than half of all the food safety risks for traded foods. Attention will be given to these two risks in this chapter.

Assessing international food safety violations is one of many risk assessment tools that can lead to food recalls (and corrections), discussed in the next chapter on food safety risk management tools. This chapter, however, emphasizes food safety performance based on more particular chemical risks, microbial risks, and national reporting on food consumption. This is followed by a discussion on inspections and audits.

Chemical Risks

Two indicators were selected to assess food safety chemical risks, one at each end of the supply chain from production to consumption. On the production side, the metric selected was the measurement of the amount of pesticides used in each of the 17 OECD countries, expressed through kilograms of active ingredients per hectare. The average rate and response scores are reported in Table 7.1 (figures on the use of pesticides are reported in Table 7.2). Results report 10 of the 17 OECD countries as progressive, including Canada. Ireland improved its response since 2010, while Belgium's pesticide use worsened. The intensive use of pesticides by Japan, the Netherlands, and Italy gave each nation a regressive score.

On the consumption side, food safety chemical risks are commonly reported through TDS. These studies provide dietary exposure results (as consumed) to chemical food contaminants such as heavy metals or trace elements, mycotoxins, radionuclides, pesticide residues, and food additives. Attempts were made to find TDS reporting in all 17 OECD countries, but TDS programs could not be found in three countries (i.e., Austria, Denmark, and Switzerland). As each nation's TDS cover variable lists of priority foods and food items, their international data are not harmonized. Therefore, the performance metric reflects the reporting frequency of each nation's TDS rather than their information quality.

Table 7.1 Ranking of rate of use of agricultural chemicals (pesticides).

Country	Average (tonnes/1000 ha)	Score	Change from 2010 (trend)[a]
Australia	1.1	Progressive	→
Austria	2.4	Progressive	→
Belgium	10.6	Regressive	↓
Canada	0.9	Progressive	→
Denmark	1.4	Progressive	→
Finland	0.7	Progressive	→
France	4.1	Moderate	→
Germany	3.1	Progressive	→
Ireland	2.3	Progressive	↑
Italy	7.6	Regressive	→
Japan	14	Regressive	→
Netherlands	9.9	Regressive	→
Norway	0.8	Progressive	→
Sweden	0.8	Progressive	→
Switzerland	3.9	Moderate	→
United Kingdom	3.5	Moderate	→
United States	2.4	Progressive	→

Sources: FAOSTAT, Active Ingredient Use in Arable Land & Permanent Crops, Tonnes per 1000 Hectares; Edition of the OECD Environmental Database for Australia and Canada.

a) Trend when compared with results in the *World Ranking: 2010 Food Safety Performance* report: → no change; ↓ worsening trend; ↑ improving trend.

The scores in Table 7.3 are thus based on the frequency of TDS reporting, public availability, and recent publication of the reports. Regressive response scores were given to countries where no TDS could be found. Progressive scores were given to TDS produced quinquennially (every 5 years) or less. Only France and Italy received such scores. Canada's older and less frequent TDS reporting earned it a moderate response score. Although other countries are to be commended for their regular reporting, they also received moderate response scores either for inadequate access to TDS results or for reporting published more than 5 years ago. Response scores are expected to improve for European nations, as many are now taking part in TDS-Exposure, which is currently pursuing harmonized TDS and sampling and analysis.

Table 7.2 Ranking based on use of agricultural chemicals.

Country	2000	2001	2002	2003	2004	2005	2006	2007	2008	2009	2010	Average	Score
Australia	0.7	0.64	1.11	1.38	1.39	1.28	1.46					1.1	Progressive
Austria	2.42	2.15	2.11	2.32	2.27	2.34	2.35	2.43	2.95	2.46	2.58	2.4	Progressive
Belgium	10.83	9.82	10.76	10.28	10.64	11.29						10.6	Regressive
Canada	0.92	0.93	0.91	0.84	0.88	0.89	0.9		1.09			0.9	Progressive
Denmark	1.39	1.44	1.26	1.3	1.27	1.39	1.37	1.43	1.71	1.15	1.6	1.4	Progressive
Finland	0.52	0.65	0.73	0.75	0.67	0.64	0.73	0.65	0.72	0.75	0.78	0.7	Progressive
France	5	5.09	4.21	3.81	3.89	3.98	3.65	3.96	4.05	3.28		4.1	Moderate
Germany	2.93	2.78	2.87	2.95	2.89	3.01	3.19	3.37	3.58	3.24	3.39	3.1	Progressive
Ireland	1.8	1.96	2.22	2.3	2.47	2.21	2.39	2.9	2.53	2.08	2.5	2.3	Progressive
Italy	7.04	6.82	8.65	8.17	7.96	8.19	7.49	8.29	7.95	5.66	7.35	7.6	Regressive
Japan	16.53	16.43	14.75	14.33	13.72	13.6	13.97	13.15	12.69	13.23	12.1	14	Regressive
Netherlands	12.06	10.05	10.18	10.14	9.15	9.32	9.49	11.04	9.78	9.04	8.75	9.9	Regressive
Norway	0.39	0.58	0.9	0.76	0.95	0.6	0.81	0.84	0.94	0.64	0.84	0.8	Progressive
Sweden	0.64	0.66	0.82	0.96	0.51	0.79	0.83	0.81	0.89	0.73	0.75	0.8	Progressive
Switzerland	3.61	3.58	3.53	3.44	3.23	3.22	3.16	4.88	4.68	5.05	4.81	3.9	Moderate
United Kingdom	3.53	3.66	3.46	3.49	3.59	3.68	3.17	3.64	3.57	3.5	2.79	3.5	Moderate
United States	2.41	2.31	2.37	2.4	2.5	2.39	2.41	2.42				2.4	Progressive

Source: FAOSTAT. Active Ingredient Use in Arable Land and Permanent Crops.

Table 7.3 Score based on reporting of food safety chemical risks through total diet studies.

Country	TDS completed	Reporting	Score
Australia	Yes	2011	Moderate
Austria	None found		Regressive
Belgium	TDS like	First in 2010	Moderate
Canada	Yes	1992–2007, one city per year, last 2007	Moderate
Denmark	None found		Regressive
Finland	Yes	2005	Moderate
France	Yes	Quinquennial post 2001, 2006, 2011–2014	Progressive
Germany	Pilot	Future TDS planned	Regressive
Ireland	Yes	2002–2003, 2005	Moderate
Italy	Yes	1980–1984, 1994–1996, 2012–2014	Progressive
Japan	Yes	Annually since 1977	Moderate
Netherlands	Yes	Once/decade, last 2003–2004	Moderate
Norway	First	First (2012–2016)	Regressive
Sweden	TDS like	1999, 2005, 2010	Moderate
Switzerland	None found		Regressive
United Kingdom	Yes	Regularly since 1966 (latest 2009)	Moderate
United States	Yes	Regularly since 1961 (latest 2008)	Moderate

Source: Adapted from The Conference Board of Canada.

Microbial Risks

In Canada, there are some four million cases of domestically acquired food-borne each year (Thomas et al., 2013). The majority of cases (90%) relate to norovirus, *Clostridium perfringens*, *Campylobacter*, and *Salmonella*.

Although acute outbreaks receive the most media attention, many outbreaks are barely noticeable in the media. For example, from July 15 to November 13, 2014, 175 confirmed cases of *Escherichia coli* were reported in Alberta (Alberta Health Services, 2014). In September 2013, 28

E. coli illness cases were reported as related to Gouda cheese in British Columbia (PHAC, 2013a). And in late 2012, 30 *E. coli* illness cases linked to shredded lettuce were reported in the Maritimes and Ontario (PHAC, 2013b). Yet these outbreaks received little national attention.

One acute outbreak case that received much more media attention in Canada and internationally, although the number of food-borne illness cases was noticeably lower, was the 2012 XL Foods Inc. outbreak that saw 18 people sickened after eating *E. coli*-tainted beef shipped from a federally inspected plant in Brooks, Alberta (CFIA, 2012). Part of the blame fell on a weak food safety culture at the plant and a relaxed attitude toward applying mandatory procedures. For example, the plant's Hazard Analysis Critical Control Point (HACCP) plan was not reviewed or updated on a regular basis, and audit reports did not highlight infractions. However, corrections have since been made, as some labeling regulation and inspection protocol changes were implemented as a result of the recall.

Table 7.4 provides a summary of the performance scores across each of the five following food-borne illnesses, namely, *Campylobacter* spp., *Salmonella* spp., *Yersinia* spp., *E. coli*, and *Listeria monocytogenes*. Performance is ranked by comparing country-to-country scores, in this case rates of food-borne illness by the number of cases per 100 000 inhabitants over a 7-year period from 2006 to 2012. Note the outbreak data is only as good as the surveillance method applied in each country. Also, only the incidences reported are taken into account; dietary patterns and geographical location, factors that influence microbial risks, are not.

The trend is thereafter evaluated as an upward or downward variable compared with the overall scores from the 2010 world food safety ranking study. Along with Canada, countries with the strongest performance include Austria, France, Ireland, Japan, the United Kingdom, and the United States, while Scandinavian countries and Germany reported the weakest performance. When compared with performances assessed in 2010, 7 of the 17 countries, including Canada, improved their score, while the Netherlands was the only country not to have maintained or improved its food-borne illness performance.

For annual figures, Tables 7.5, 7.6, 7.7, 7.8, and 7.9 provide a numbers-based measurement of actual reported food-borne illness cases per 100 000 inhabitants. Two different sources for Canada were identified and reported: the Public Health Agency of Canada's notifiable diseases (PHAC, 2014a) and the National Enteric Surveillance Program (NESP) (PHAC, 2014b). Both provided similar results except for *Campylobacter* spp.; reported NESP results are quite low. Additionally,

Table 7.4 Ranking of incidences of reported illness by food-borne pathogen per 100 000 people.

Country	*Campylobacter* spp.	*Salmonella* spp.	*Yersinia* spp.	*Escherichia coli*	*Listeria monocytogenes*	Overall score	Change from 2010 (trend)[a]
Australia	Poor	Poor		Superior	Superior	Average	↑
Austria	Average	Poor	Superior	Superior	Superior	Superior	↑
Belgium	Average	Poor	Average	Superior	Average	Average	→
Canada	Superior	Superior	Average	Average	Superior	Superior	↑
Denmark	Poor	Poor	Poor	Poor	Poor	Poor	→
Finland	Poor	Poor	Poor	Superior	Poor	Poor	→
France	Superior	Superior	Superior	Superior	Average	Superior	→
Germany	Poor	Poor	Poor	Poor	Average	Poor	→
Ireland	Average	Superior	Superior	Poor	Superior	Superior	→
Italy	Superior	Superior	Superior	Superior	Superior	Poor[b]	→
Japan	Superior	Superior		Superior		Superior	↑
Netherlands	Average	Superior		Poor	Superior	Average	↓
Norway	Poor	Poor	Superior	Superior	Average	Average	→
Sweden	Poor	Poor	Poor	Poor	Poor	Poor	→
Switzerland	Poor	Superior		Superior	Poor	Average	↑
United Kingdom	Poor	Superior	Superior	Average	Superior	Superior	↑
United States	Superior	Superior	Superior	Superior	Superior	Superior	↑

a) Trend when compared with results in the *World Ranking: 2010 Food Safety Performance* report: → no change; ↓ worsening trend; ↑ improving trend.
b) Italy maintains its poor overall score from 2008 due to suggested underestimation of food-borne illness cases.

Table 7.5 Food-borne illness cases per 100 000 inhabitants: *Campylobacter* spp.

Country	2006	2007	2008	2009	2010	2011	2012	Average	Score
Australia	111.1	119.9	107.3	108.4	112.5	115.7		112.5	Poor
Austria	60.7	70.1	51.4	18.14	52.6	16	55.79	46.4	Average
Belgium	54.9	55.8	47.9	53.41	27.96	70.46		51.7	Average
Canada	30.64	29.11	28.46	25.85	26.55	27.63	29.3	28.2	Superior
Canada (NESP)	5.99	5.93	4.83	5.17	5.36	5.6		5.5	
Denmark	59.7	71	63.4	60.84	72.94	73.01	66.66	66.8	Poor
Finland	65.4	77.8	84	76.04	73.7	79.29	78.7	76.4	Poor
France	4.2	4.8	5.4	6.15	6.68	8.51	38.89	10.7	Superior
Germany	63.1	80.3	78.7	76.01	76.59	86.62	76.54	76.8	Poor
Ireland	43	43.7	39.8	40.67	37.15	54.3	52.17	44.4	Average
Italy		1.1	0.4	0.88	0.76	0.77	1.27	0.9	Superior
Japan	1.81	1.89	2.42					2.1	Superior
Netherlands	19.5	38.6	39.2	43.62	46.21	50.89	48.83	41	Average
Norway	55.8	60.6	60.7	59.34	55.21	61.07	58.83	58.8	Poor
Sweden	67.2	78	83.8	77.55	85.66	87.24	83.32	80.4	Poor
Switzerland		79.5	102.3	105.9	85.05	100.8	105.49	96.5	Poor
United Kingdom	86.3	95	90.9	106.32	113.37	115.44	117.43	103.5	Poor
United States	12.73	12.81	12.64	12.96	13.52	14.31	14.3	13.3	Superior

Table 7.6 Food-borne illness cases per 100 000 inhabitants: *Salmonella* spp.

Country	2006	2007	2008	2009	2010	2011	2012	Average	Rank
Australia	39.7	44.9	38.6	43.3	53.5	54.2		45.7	Poor
Austria	57.9	40.7	27.7	33.2	26	17.1	21	31.9	Poor
Belgium		37.5	35.9	29.2	29.2	29		32.2	Poor
Canada	16.82	18.69	18.27	17.44	20.6	19.23	19.67	18.7	Superior
Canada (NESP)	17.51	19.42	18.99	17.97	21.17	19.68		19.1	
Denmark	30.6	30.5	67	38.6	29.1	21	21.6	34.1	Poor
Finland	49	51.9	59	43.7	45.3	38.7	40.8	46.9	Poor
France	10.1	8.7	11.3	11.1	11.1	13.4	13.3	11.3	Superior
Germany	63.8	67.3	52.2	38.3	30.4	29.3	25.1	43.8	Poor
Ireland	9.9	10.2	10.2	7.5	7.8	6.9	6.7	8.5	Superior
Italy	8.8	7.6	5.4	6.9	4.5	5.5		6.5	Superior
Japan	1.62	2.84	2.01					2.9	Superior
Netherlands	10.2	11.9	15.5	11.4	13.6	12	20.5	13.6	Superior
Norway	39.1	35.2	41	25.7	25.7	26.2	27.5	31.5	Poor
Sweden	44.8	43.1	45.6	33	38.7	30.7	30.8	38.1	Poor
Switzerland	23.9	23.7	26.6	17.2	15.1	16.4	16.1	19.9	Superior
United Kingdom	23.3	22.7	18.8	17	15.6	15.1	14.3	18.1	Superior
United States	14.76	14.89	16.09	15.02	17.55	16.47	16.42	15.9	Superior

Table 7.7 Food-borne illness cases per 100 000 inhabitants: *Yersinia* spp.

Country	2006	2007	2008	2009	2010	2011	2012	Average	Rank
Australia									
Austria	1.9	1.7	1.1	1.68	1	1.42		1.5	Superior
Belgium	2.5	2.3	2.6	2.23	1.99	1.95		2.3	Average
Canada			2	1.5	4.3	3.25		2.8	Average
Canada (NESP)	1.77	1.48	1.24	1.13	1	1.1		1.3	Superior
Denmark	4	5	6	4.32	3.49	4.05		4.5	Poor
Finland	15.1	9.1	11.5	11.88	9.75	10.31		11.3	Poor
France			0.3	0.32	0.37	0.45		0.4	Superior
Germany	6.3	6.1	5.3	4.51	4.09	4.14		5.1	Poor
Ireland	0.1	0.1	0.1	0.07	0.07	0.13		0.1	Superior
Italy				0.02	0.02	0.02		0	Superior
Japan									
Netherlands									
Norway	1.9	1.5	1.1	1.25	1.07	1.22		1.3	Superior
Sweden	6.2	6.2	5.9	4.29	3.01	3.72		4.9	Poor
Switzerland									
United Kingdom	0.1	0.1	0.1	0.1	0.09	0.09		0.1	Superior
United States	0.36	0.36	0.36	0.33	0.34	0.34	0.33	0.3	Superior

Note: In Canada, *Yersinia* is not a nationally notifiable disease.

Table 7.8 Foodborne illness cases per 100 000 inhabitants: *E. coli* VTEC/STEC.

Country	2006	2007	2008	2009	2010	2011	2012	Average	Rank
Australia	1	0.3	0.5	0.5	0.6	0.4	0.4	0.5	Superior
Austria	0.5	1	0.8	1.09	1.05	1.43	1.54	1.1	Superior
Belgium	0.4	0.4	1	0.9	0.77	0.91	0.95	0.8	Superior
Canada	3.31	3.24	2.29	1.82	1.6	1.87	1.94	2.3	Average
Canada (NESP)	2.99	2.83	1.98	1.56	1.18	1.39		2	
Denmark	2.7	2.9	2.9	2.9	3.16	3.87	3.46	3.1	Poor
Finland	0.3	0.2	0.2	0.54	0.39	0.5	0.56	0.4	Superior
France	0.1	0.1	0.1	0.14	0.16	0.34	0.32	0.2	Superior
Germany	1.4	1.1	1.1	1.07	1.59	6.8	1.93	2.1	Poor
Ireland	3.6	2.7	4.8	5.33	4.41	6.14	8.99	5.1	Poor
Italy	0.1	0.1	0.1	0.08	0.05	0.08	0.08	0.1	Superior
Japan	0.85	1.24	0.49					1	Superior
Netherlands	0.3	0.5	0.6	1.9	2.88	5.07	6.27	2.5	Poor
Norway	1.1	0.6	0.5	2.25	1.03	0.96	1.5	1.1	Superior
Sweden	2.9	2.9	3.3	2.46	3.58	4.96	4.98	3.6	Poor
Switzerland	0.6	0.7	0.9	0.54	0.4	0.9	0.79	0.7	Superior
United Kingdom	2.1	1.9	1.9	2.19	1.79	2.41	2.17	2.1	Average
United States	1.3	1.2	1.12	0.99	0.95	0.98	1.12	1.1	Superior

Table 7.9 Food-borne illness cases per 100 000 inhabitants: *Listeria monocytogenes*.

Country	2006	2007	2008	2009	2010	2011	2012	Average	Rank
Australia	1	0.3	0	0.3	0.4	0.3	0.3	0.4	Superior
Austria	0.1	0.2	0.4	0.55	0.41	0.31	0.43	0.3	Superior
Belgium	0.6	0.5	0.6	0.54	0.37	0.64	0.75	0.6	Average
Canada		0.39	0.64	0.37	0.44	0.41	0.36	0.4	Superior
Canada (NESP)						0.38		0.4	
Denmark	1	1.1	0.9	1.76	1.12	0.88	0.9	1.1	Poor
Finland	0.9	0.8	0.8	0.64	1.33	0.8	1.13	0.9	Poor
France	0.5	0.5	0.4	0.51	0.48	0.43	0.53	0.5	Average
Germany	0.6	0.4	0.4	0.47	0.46	0.4	0.5	0.5	Average
Ireland	0.2	0.5	0.3	0.22	0.22	0.16	0.24	0.3	Superior
Italy	0.1	0.1	0.1	0.15	0.16	0.14		0.1	Superior
Japan									
Netherlands	0.4	0.4	0.3	0.27	0.43	0.52	0.44	0.4	Superior
Norway	0.6	1	0.7	0.65	0.47	0.43	0.6	0.6	Average
Sweden	0.5	0.6	0.7	0.79	0.67	0.59	0.76	0.7	Poor
Switzerland	1		0.6	0.53	0.9	0.7	0.5	0.7	Poor
United Kingdom	0.3	0.4	0.3	0.38	0.28	0.26	0.3	0.3	Superior
United States	0.28	0.26	0.26	0.32	0.27	0.28	0.25	0.3	Superior

results for Italy are overly spectacular: They show that food-borne illnesses are practically nonexistent, suggesting the values are underestimated. Consequently, Italy maintains its poor overall outcome score as per the initial score from the 2008 world food safety ranking study. Further investigation is required to review Italy's reporting methods or to emulate the country's best practices if its incidence reporting is indeed correct. A future metric could also review the food safety regulations in different countries and then rank these with respect to controlling pathogens. For example, the American zero tolerance policy on *Listeria* has likely contributed to a larger number of product recalls, requiring additional sampling and testing, even for foods that do not support *Listeria* growth and should be considered a lesser public health concern.

National Food Consumption Reporting

Nutrition surveys, food consumption surveys, dietary exposure assessments such as TDS, and intake assessments are essential parts of assessing exposure to food safety risks. They not only help inform food safety recommendations but also influence nutrient-based recommendations, food-based dietary guidelines, and labeling and reference values, among other benefits. A country's regulatory agency can then provide food safety information, through food labeling and other means, to allow people to recognize and avoid foods that present a risk such as an allergenic risk.

Each nation pursues a range of tools to assess consumption, such as nutrition surveys, health surveys with secondary questions regarding intake or nutrition, and household budget surveys. This benchmarking assessment does not address the quality of the data or compare the methods used. Similar to the TDS metric, the food consumption indicator in this study reflects the latest reporting and frequency of national food or nutrition intake surveys. Superior outcome scores were attributed to recent and frequent surveys that were less than 7 years old. Question marks reflect estimated frequencies. Table 7.10 indicates that seven countries reported superior outcome scores, while poor scores were attributed to four countries with nutrition surveys over a decade old. Canada received an average score for its lower frequency. As many surveys are over a decade old, this report calls for a greater frequency of national food consumption or dietary intake surveys to assess the exposure to food safety risks, among other benefits.

Table 7.10 Score for national food consumption or dietary intake reporting.

Country	Latest year	Frequency (years)	Score
Australia	National Nutrition Survey 1995		Poor
	Australia's Food and Nutrition 2012	2014	
	(2014–2015 National Health Survey to include basic measures of nutrition)	Based on 1995	
Austria	Austrian Study on Nutritional Status 2010–2012	4	Superior
Belgium	Enquête de consommation alimentaire 2014 (previous 2004)	10	Average
Canada	2004 CCHS Nutrition Cycle (next 2015)	Once per decade	Average
	Last national food consumption study 1970–1972. CCHS 2001, 2003, 2005, 2007–2008, then rotating annual collection 2-year cycle		
Denmark	Danish National Survey of Dietary Habits and Physical Activity 2011–2013	Once per decade	Average
Finland	FINDIET 2012	5	Superior
France	Individual/National Study on Food Consumption (INCA) 2006–2007	7	Superior
	(next 2014)		
Germany	National Food Consumption Study II 2005–2006	2015	Poor
	National Nutrition Monitoring from NVS II	Annual since 2008	
Ireland	North/South Ireland Food Consumption Survey 2000	No frequency established	Poor
	National Adult Nutrition Survey 2011		
	National Teens' Food Survey 2005–2006		
	National Children's Food Survey 2003–2004		
Italy	3rd National Food Consumption Survey, INRAN-SCAI 2005–2006	Once per decade	Average
	(previous 1994–1996 INN-CA Study)		

(*Continued*)

Table 7.10 (Continued)

Country	Latest year	Frequency (years)	Score
Japan	National Health and Nutrition Survey 2011 Data availability for 2007 survey Dietary Reference Intakes (DRIs) revised every 5 years	Annually since 1948	Superior
Netherlands	Dutch National Food Consumption Survey—Core 2012–2016 DNFCS—Young adults (2003) DNFCS—Young children (2005–2006) DNFCS—Core survey 7–69 years (2007–2010)	5 (core)	Superior
Norway	Norkost 3 (National Dietary Survey) 2010–2011 Two national diet surveys of adults were conducted: 1993–1994 (Norkost 1) and 1997 (Norkost 2)	2014	Average
Sweden	Riksmaten (Dietary Survey) Adults 2010–2011 Riksmaten children 2003 Riksmaten 1997–1998 dietary habits and nutrient intake	2014	Average
Switzerland	National Nutrition Survey menuCH 2014–15	No previous study found	Poor
United Kingdom	National Diet and Nutrition Survey Rolling Program (2008/2009–2010/2011) Food Statistics Pocketbook (2013)	Annual rolling basis since 2008	Superior
United States	National Health and Nutrition Examination Survey 2012 (current 2013–14)	Continuous program since 1999	Superior

Source: Adapted from The Conference Board of Canada.

Table 7.11 Nutrition information quality scores for selected EU countries.

Country	Score
Austria	Progressive
Belgium	Progressive
Denmark	Progressive
Finland	Progressive
France	Progressive
Germany	Progressive
Ireland	Moderate
Italy	Moderate
Netherlands	Progressive
Norway	Moderate
Sweden	Moderate
United Kingdom	Progressive

Source: García-Alvarez et al. (2009). Reproduced with permission of Cambridge University Press.

Food safety risk perception is vital when assessing how consumers potentially react to certain immediate threats or the response to threats, such as a major food recall. For governments, improving data collection and quality in the area of food safety and risk perception is critical moving forward. In terms of data quality, a scoring system to rate and compare the quality of surveys for nutrient intake across Europe was developed in 2009 to assess best practices (Blanquer et al., 2009). Authors used both systematic literature and government website searches, similar to the approach used in this study, followed by 2–5 questionnaires that were sent to nutrition experts in each country to fill information gaps and validate findings. Indeed, the use of questionnaires is an additional research instrument to be considered for any ulterior global food safety ranking study.

Table 7.11 provides the comparative response scores for nutrition information quality assessed in the 2009 European study. No European OECD country ranked poorly in the study's scoring system. However, four countries—Ireland, Italy, Norway, and Sweden—received moderate scores, indicating that the quality of their nutrition information requires additional attention.

Inspections and Audits

Previous world food safety rankings from 2008 and 2010 assessed the rates of inspections and audits, measured by whether a country had strict risk-based inspection policies and carried out frequent inspections (e.g., daily or weekly). Data considered for this 2014 comparative study found that countries do regularly monitor and inspect foods for safety, from municipal to federal/national levels, but that the benchmarking choices discussed earlier are insufficient to reliably assess their food safety performance. Indeed, more objective indicators are needed to satisfactorily assess food safety performance expressed through inspections and audits. Performance could be determined, for instance, through the annual number of inspection/audits per food, feed, or commodity business operator (for export and local consumption). It would not, however, assess the accuracy of the inspections.

Food safety performance might also be assessed by examining the total number of inspectors or inspections in absolute terms, though this may be a less relevant measure of food safety performance. For instance, Canada had 3577 federal field inspection staff in 2013, higher than many OECD countries. However, the number itself does not speak for food safety performance or correspond to increased food safety standards.

Ratios or food safety performance in relative terms is more meaningful and could be assessed, for example, through the number of audits per inspector (not the number of inspectors or audits), or compliance rates, or the number of veterinary/fishery inspections per premise. The result or findings from inspection are also useful in measuring food safety performance, rather than just the number of inspections. Another possible food safety metric could include average HACCP limit deviations, for example, or the percentage of audits with nonconformities (e.g., 49% in Sweden vs. 7.5% in Denmark in 2007 (IFC, 2008)).

Attempts were made to collect information and report metrics in relative terms here. However, data availability for these metrics was limited, difficult to acquire, or unavailable to the public. Any future ranking study should consider investing in developing these indicators further which could be acquired through surveys sent to food safety experts in each OECD country.

Food Safety Risk Management

Managing food safety risks is both a policy-based and a commercially based process to prevent, control, and mitigate risks while ensuring health protection and fair trade practices. Managing food safety risks requires infrastructure such as food safety standards (public and private); laws, regulations, and policies that facilitate food safety controls; regular enforcement and surveillance; timely, effective emergency response mechanisms, including traceability and food recalls; import controls; and risk-based food safety monitoring and laboratory capacity and services. It also requires public and private systems to nurture a culture that puts a deliberate emphasis on food safety practices on a daily basis. For the purposes of this benchmarking study, the following metrics were selected: national food safety response capacity, food recalls, food traceability, and radionuclide standards.

National Food Safety Response Capacity

A global search for existing scoring of food safety system response capacity and infrastructure led to the reporting by World Health Organization (WHO) states parties (countries) regarding their progress in implementing the 2005 International Health Regulations (IHR) (WHO, 2013). The capacity scores describe current national capacity monitoring for different categories related to legislation, coordination, surveillance, response, preparedness, risk communication, laboratory capacity, points of entry, and other categories.

Among these categories, specific areas of concern were added, such as food safety, zoonoses, and chemical emergencies that may constitute a public health event of international concern. The food safety concern more specifically addresses the mechanisms established and functioning for detecting and responding to food-borne disease and food contamination.

Table 7.12 lists the results from the self-reporting each nation provides to the WHO. For food safety, for instance, there are 18 questions and criteria based upon core capacities that are foundational (food standards, risk identification, response expertise) and either input based (e.g., inspection), output based (e.g., laboratory capacity), or achievement based (e.g., published analysis and results). A copy of the 18 questions can be found in Table 7.13. A score of 100 means that the states party has met all 18 criteria. As the IHR food safety core capability focuses on food-borne disease and food contamination, the benchmark evaluation for ranking countries according to capacity includes zoonoses and chemical core capabilities. The assessment reveals that in 2012 and 2013, 11 of 17 OECD countries, including Canada, had superior outcome scores, while four countries still did not meet all of the food safety required criteria.

Table 7.12 Outcome score for national capacity to respond to food safety and other emergencies.

Country	Food safety (food-borne illness)	Zoonotic (animal)	Chemical emergencies	Score
Australia	87	100	100	Superior
Austria	93	89	46	Poor
Belgium	100	89	92	Average
Canada	100	100	100	Superior
Denmark	100	89	100	Superior
Finland	87	100	77	Average
France	100	100	85	Superior
Germany	100	100	100	Superior
Ireland	100	100	92	Superior
Italy	N/A	N/A	N/A	
Japan	100	100	100	Superior
Netherlands	100	100	100	Superior
Norway	100	100	85	Superior
Sweden	87	100	85	Average
Switzerland	100	100	100	Superior
United Kingdom	N/A	N/A	N/A	
United States	100	100	100	Superior

Source: Global Health Observatory Data Repository 2012 and 2013. Reproduced with permission of WHO.

Table 7.13 Food safety core capability questions and criteria in the World Health Organization's International Health Regulations National Capacity Monitoring Survey.

1	Are national or international food safety standards available?
2	Are there national food laws, regulations or policies in place to facilitate food safety control?
3a	Are national food laws, regulations or policies up to date?
3b	Are national food laws, regulations or policies implemented?
4	Has a coordination mechanism been established between the food safety authorities, for example, the INFOSAN Emergency Contact Point (if member) and the IHR national focal point?
5	Are there functional mechanisms in place for multisectoral collaborations for food safety events?

Table 7.13 (Continued)

6	Is your country an active member of the INFOSAN network?
7	Is a list of priority food safety risks available?
8	Are guidelines or manuals on the surveillance, assessment, and management of priority food safety events available?
9	Have the guidelines or manuals on the surveillance, assessment, and management of priority food safety events been implemented?
10	Have surveillance, assessment, and management of priority food safety events been evaluated and relevant procedures updated as needed?
11	Is epidemiological data related to food contamination systematically collected and analyzed?
12	Are there risk-based food inspection services in place?
13	Does the country have access to laboratory capacity (through established procedures) to confirm priority food safety events of national or international concern including molecular techniques?
14	Is there timely and systematic information exchange between food safety authorities, surveillance units, and other relevant sectors regarding food safety events?
15	Is there a roster of food safety experts for the assessment and response to food safety events?
16	Have operational plan(s) for responding to food safety events been implemented?
17a	Have operational plan(s) for responding to food safety events been tested in an actual emergency or simulation exercise?
17b	Have operational plan(s) for responding to food safety events been updated as needed?
18	Have mechanisms been established to trace, recall, and dispose of contaminated products?

Source: WHO International Health Regulations (2005). Reproduced with permission of WHO.

Food Recalls

Food recalls are a vital part of food safety management infrastructure and response capacity. Recalls are the means by which the food industry and government food regulators ensure food safety by removing food products from the supply chain, inventories, points of sale, store shelves, and even households. Food recalls may be either voluntary or mandatory and fall into one of three classes: for instance, in Canada, the United States, and elsewhere, class I (high risk), class II (moderate risk), or class III (low and no risk).

From April 2012 to March 2013, Canada experienced 268 food recalls, or 7.6 food recalls per million Canadians. Over this period, recalls were most commonly due to undeclared allergens (109 recalls, or 41% of all recalls) and microbial contamination (81 recalls, or 30%). Elsewhere, on average, according to AIG, "30 class I and II product recalls occur every week in the food and beverage industry in the United States, and another 22 equivalent recalls are reported in Europe" (AIG, 2014).

Interestingly, according to a review of the chemical, biological, and operational hazards from 2004 to 2010 within the US, the UK, and Irish agri-food industries, operational product recalls were the most frequent (55% of all recalls) (Potter et al., 2012). These included product or packaging defects, product contamination, mislabeling, and unauthorized ingredients. Biological hazards came in second (36%), followed by chemical hazards (9%). The study also provides the frequency of product recalls by subindustry and the distribution of industry-specific product recalls, for example, the largest number of product recalls stem from the processed food industry at 24%.

Recalls are thus common and vary in severity and by industry. The number of recalls themselves is certainly significant. Also of interest is the level of surveillance. Of greater significance for performance, however, is the ratio of food recalls normalized to each nation's population. This food safety performance metric provides a more relative and comparative food safety performance measure. It can be argued that the higher the ratio, the stronger the performance and evidence of the strength of the country's food safety system. Indeed, consumers feel more confident when there is a food recall, as it shows the system works. Confidence, however, does not correlate with actual performance in preventing unsafe foods from entering the food market.

More significantly, the data also allow for comparisons from 1 year to the next to be more valid than they have been historically. Since most food safety regulators have been in existence for some time, a reduction or an increase in the number of recalls may not be desirable. In essence, this study assesses the level of stability in the number of food recalls. Over time, as food safety systems mature around the world, we would expect to see the number of recalls generally stabilize as it has in recent years. Thus, if the number of recalls in 2013 fell within the historical range of a country, the data would suggest that 2013 was a normal year. Consequently, benchmarking of national food recalls across OECD countries was appraised through comparing each country's performance in 2013 with the median number of food recalls over a 4-year period from 2009 to 2012. Country scores in Table 7.14 were calculated by comparing the variations from their respective median number of food recalls per million inhabitants.

Table 7.14 Response (consistency) score for food recalls per 1 000 000 people.

Country	2009	2010	2011	2012	Range	Median	2013	Variation from median (%)	Score
Australia	2.5	2.4	3	2.6	2.4–3.0	2.6	1.8	−30.8	Regressive
Austria	13.1	10.5	7.7	5.8	5.8–13.1	9.1	5.4	−40.6	Regressive
Belgium	10.8	8.6	11.6	12.9	8.6–12.9	11.2	14.6	30.3	Regressive
Canada	7	6.2	7.7	8.7	6.2–8.7	7.4	7.6	2.7	Progressive
Denmark	22.1	23.6	27.1	23.2	22.1–27.1	23.4	20	−14.5	Moderate
Finland	26.4	24.2	20.6	19.4	19.4–26.4	22.4	16.3	−27.2	Regressive
France	2.4	2.6	3	4.2	2.4–4.2	2.8	3.9	39.2	Regressive
Germany	5	4.8	5.1	4.5	4.5–5.1	4.9	4.1	−16.3	Moderate
Ireland	6.6	7.2	10.7	11.6	6.6–11.6	9	8.7	−3.3	Progressive
Italy	7.7	8.9	9	8.7	7.7–9	8.8	8.9	1.1	Progressive
Japan					N/A				Regressive*
Netherlands	12.8	12.9	12.1	10.3	10.3–12.9	12.5	15.7	25.6	Regressive
Norway	6.2	4.7	10.3	12.2	4.7–12.2	8.3	8.8	6.0	Moderate
Sweden	6.5	7.8	7.6	10	6.5–10	7.7	9.5	23.3	Regressive
Switzerland	0.5	0.9	0.8	2.5	0.5–2.5	0.8	4.9	512.5	Regressive
United Kingdom	5.4	5.1	8	8.1	5.1–8.1	6.7	5.1	−23.8	Regressive
United States	1.7	2.4	1.6	2.2	1.6–2.4	2	1.9	−5.0	Moderate

Sources: National population figures taken from The World Bank.

* Japan is listed as regressive as numbers could not be readily found or were not available.

Countries were thus rewarded for their consistency. Canada, Ireland, and Italy received a progressive score, while several large variations from the medians were common and of concern in 2013, with the most disquieting performances going to Switzerland and France. The year 2013 was also atypical for Belgium, the Netherlands, and Sweden. All these countries reported a higher number of recalls than their median over four previous years would suggest. No recall data could be found for Japan, which was given a regressive grade.

Food Traceability

Food traceability is a product tracing tool to identify sourcing across the supply chain, at least one step back (whence it came) and one step forward (down the supply chain). Traceability is also "the ability to access any or all information relating to that which is under consideration, throughout its entire life cycle, by means of recorded identifications" (Olsen and Borit, 2013). Traceability systems therefore help to manage food safety risks and authenticate source and compliance. National food traceability systems can also improve market access and supply chain coordination. They have generally focused on livestock and animal production, though traceability is increasingly common in other sectors.

Traceability has progressively gone mainstream. Consumers are increasingly better educated about what the food industry can and cannot do to mitigate food safety risks. Innovative technologies have placed traceability systems within financial reach of most companies in the industry. Ethnicity, health benefits, fair trade, organic foods, gluten-free food, and labels—new products are continually being launched to serve very well-defined markets. Modern consumers are expecting the food industry to increase its efforts to cater to a fragmented market. Traceability can also strengthen trade options, improve confidence, reduce restrictions on exports, and mitigate the need for regulations (Charlebois et al., 2014a).

Previous traceability rankings in 2008 and 2010 investigated only one outcome measure, the depth or comprehensiveness of each nation's food traceability system. A more recent 2014 study looked at traceability across 10 metrics, across national food tracking systems of similar OECD countries. Table 7.15 depicts the study's overall score across the study's 10 traceability performance metrics. Results clearly show European nations' leading traceability performance, while other nations outside Europe trail behind. Efforts to improve performance include Japan's rice traceability regulations and Canada's progress on livestock and swine identification. However, Canada has no national supply chain

Table 7.15 Aggregate outcome scores for food traceability.

Country	Score
Australia	Average
Austria	Superior
Belgium	Superior
Canada	Average
Denmark	Superior
Finland	Superior
France	Superior
Germany	Superior
Ireland	Superior
Italy	Superior
Japan	Average
Netherlands	Superior
Norway	Superior
Sweden	Superior
Switzerland	Superior
United Kingdom	Superior
United States	Average

Source: Charlebois et al. (2014b). Reproduced with permission of John Wiley and Sons.

traceability regulations in place, notably for commodities and products outside animal production. Overall, to improve food traceability system performance globally, nations and markets should work toward uniform requirements for electronic identification, database programs, recognized identification standards, and auditing schemes (Charlebois et al., 2014a).

Radionuclide Standards

There is greater public awareness now than ever before about radiation in food (expressed through radionuclide standards), following the Fukushima Daiichi nuclear power plant accident of 2011. Radiation is measured in becquerel (Bq), while radiation in food is expressed as becquerel per kilogram (Bq/kg) or per liter (Bq/l). Table 7.16 lists the limits for radioactive cesium in foodstuffs, a common radiation measurement in food samples.

Table 7.16 Standard limits for radioactive cesium by food group.

Limits for radioactive cesium	Drinking water (Bq/l)	Milk (Bq/l)	General foodstuffs (Bq/kg)	Food items for babies (Bq/kg)	Score
Japan (new statutory limits from April 2012)	10	50	100 (plus dairy products)	50	Progressive
Australia (operational interventional limits in a reactor accident)	300	300	200	200	Progressive
EU	600	370	600	600	Progressive
Canada	100	300	1000	1000	Moderate
United States	1200	1200	1200	1200	Regressive
International (Codex)	1000	1000	1000	1000	Moderate

Sources: Carpenter and Tinker (2012), 16; Codex Alimentarius (1995), 33; Health Canada (2000).

In April 2012, the Japanese instituted new statutory limits for radioactive cesium, by far the most progressive response in the world. Limits for radionuclides in foodstuffs during a nuclear emergency can be lowered and are also considered progressive, as in Australia's case. European Union (EU) standards were also awarded a progressive response score because they are lower than the international (*Codex Alimentarius*) food standard guideline level of 1000 Bq/kg as applied to foods for human consumption and traded internationally. While Canada follows this international standard for general foodstuffs and baby foods, the American standard received a regressive score for its standard limit of 1200 Bq/kg and Bq/l.

Many countries tested imported Japanese foods for radiation to minimize the impact on the health of consumers. Studies and data in Australia (Carpenter and Tinker, 2012) and Canada (Health Canada, 2012) found such impacts to be negligible. However, in Japan, the average radiation level (cesium-134 and cesium-137) in contaminated food analyzed in December 2013 was still high at 39.71 Bq/kg (18.42 Bq/kg in fish products) and quite worrisome in wild boar meat at 6000 Bq/kg.

Significantly, while many of these foods surpassed the Japanese legal limit, they remained below international standards and are acceptable for consumption elsewhere outside Japan. Thus, it may be judicious to review the impact and exposure of radiation levels on human health and therefore reduce the international standard limits following Japan's example, given that many countries, such as Canada, the United States, and several European countries, also have nuclear reactors. Ideally, the aim is to prevent exposure, reduce exposure to nil, and maintain a lower harmonized, uniform nonemergency limit, similar to emergency situations, particularly for traded commodities.

Food Safety Risk Communication

Food crises are increasingly being considered in food safety system management planning. Consumers are also gradually becoming accustomed to more risk mitigation communications by the food industry and national governments, which are also responsible for food recalls.

The breadth of food recalls has encouraged risk communication systems to become more strategic and institutionalized within government and the food industry. Yet, some institutionalized food safety risk communication systems falter and warrant adjustments. Thus, given how food crises can differ, all risk communication strategies need to be adept and nimble from one incident to the next.

The exchange of information and opinions around food safety risks and incidents (actual or perceived) is also part of risk communication. It is the responsibility of food safety stakeholders to communicate and coordinate information among themselves, but they must also relate information to the consumer and foster public trust or confidence in how food safety risks are assessed, managed, explained, and corrected.

Several potential metrics were considered to assess food safety risk communication performance. Metrics such as food safety education programs are deemed part of risk communication (and risk management thereafter once integrated and implemented). Food safety education and training, such as safe food handling and practices in the home and in the food service and processing industry, are large areas of risk and where outbreaks occur even if they are not perceived as such. Moreover, poor food safety handling practices can negate performance gains and food safety assurances built in earlier in the supply chain. However, for the purposes of this study, no objective or outcome metric could be found that reported nationally comparable results with regard to completed food safety training, changes in handling practices, food safety in school curricula, and so on. Indeed, any subsequent global ranking study should consider the development of survey instruments to gather adequate comparable evidence on food safety education, practices, and training.

Knowing who communicates risks to the public is also important, as is the existence of risk management plans, recall regulations, and participation in food safety alert systems. But these plans and tools do not describe performance. Once more, no objective metric could be identified to assess risk communication strategies and infrastructure or governance. Therefore, the benchmarking assessment in this chapter relates to the consumer. Two metrics were selected: allergenic risks, often the most common cause of food recalls, and public trust, or a measure of confidence, as it provides the food safety system with a consumer's perspective of performance and possibly even their understanding of the food safety system.

Allergenic Risks and Labeling

Food labeling and the indication of potential allergens are simple but effective means of communicating food safety risks to stakeholders across the supply chain. They serve to inform and warn the consumer at the point of purchase of food ingredients that can cause an allergic reaction. In Canada, the self-reported prevalence of a food allergy is 6.67%, though diagnosed prevalence levels are generally lower in westernized countries and are estimated at 5–8% in young children and 3–4% in adults (Soller et al., 2012).

The benchmarking of allergenic risk performance through food labeling practices is thus a valuable metric to compare national food safety performances and foster accountability. A 2013 study looked at nine food allergenic risk metrics required for labeling and indication of allergens. These include the name of the food item, use by or made on dates, the rigor of nutritional information, the list of ingredients, the country of origin, and the indication of food additives, as well as directions for use and storage, the dealer's name and address, and lot/batch identification. The aggregate results are shown in Table 7.17. Five countries, including Canada, received a progressive response score for combined results above six, the base score for moderate performance. Australia and Switzerland had the best results, while many metrics are common to all countries, including the name of the food, dates, the list of ingredients, and the indication of food additives. Only five countries provide nutritional information and country-of-origin information. However, Japan and the United States do not provide lot/batch identification. Canada did not receive full marks, as it had not met the criteria for direction for use and storage.

Table 7.17 Responsive score for labelling food and indicating allergens.

Country	Result	Gaps	Score
Australia	9		Progressive
Austria	6	Nutritional information Country of origin Dealer name and address	Moderate
Belgium	6	Nutritional information Country of origin Dealer name and address	Moderate
Canada	8	Direction for use and storage	Progressive
Denmark	6	Nutritional information Country of origin Dealer name and address	Moderate
Finland	6	Nutritional information Country of origin Dealer name and address	Moderate
France	6	Nutritional information Country of origin Dealer name and address	Moderate
Germany	6	Nutritional information Country of origin Dealer name and address	Moderate
Ireland	6	Nutritional information Country of origin Dealer name and address	Moderate
Italy	6	Nutritional information Country of origin Dealer name and address	Moderate
Japan	7	Nutritional information Lot/batch identification	Progressive
Netherlands	6	Nutritional information Country of origin Dealer name and address	Moderate
Norway	6	Nutritional information Country of origin Dealer name and address	Moderate
Sweden	6	Nutritional information Country of origin Dealer name and address	Moderate
Switzerland	9		Progressive
United Kingdom	6	Nutritional information Country of origin Dealer name and address	Moderate
United States	7	"Use by" or "made on" dates Lot/batch identification	Progressive

Source: Balazic et al. (2013).

Public Trust

International food safety incidents receive considerable media attention. Recent examples include the Canadian listeriosis outbreak in 2008, radioactive beef and other foods from Japan (2011), *E. coli*-tainted Egyptian fenugreek seeds in Europe (2011), traces of horsemeat in beef products in the United Kingdom and Europe (2013), the organic egg fraud in Germany (2013), and various repeated concerns in China, including tainted milk, baby food, meats, and other products. But how often do consumers worry about food safety in general? A recent 2014 study showed most Italians worry (80%), followed by French (61%), German (49%), and English (36%) consumers (Friends of Glass, 2014). In Germany, for instance, 48% of survey respondents suspect that significant aspects of ingredients are hidden or absent altogether (SGS, 2010).

Domestically, in the case of Canada's largest food recall in history tied to *E. coli* cases in ground beef and beef trim in 2012 in Alberta, Canadians maintained their trust in the safety of ground beef (Charlebois et al., 2014b). Moreover, Canadian consumer food safety concerns typically rise following international food safety incidents. As Canadians consider food from Canada safer than from abroad, they increase their purchases of local or Canadian food following international incidents (BDBC, 2013). Similar results were found following the 2006 spinach recall and 2008 processed meat recall (Charlebois and Watson, 2009).

Although consumers consider food to be safe, they remain concerned about food contaminants and food-borne illnesses in particular. Public trust or confidence therefore provides the food safety system with a consumer's perspective of performance and possibly an understanding of food safety systems. However, surveys of trust for a given year are less valuable as a performance metric than changes in trust measured over time. For instance, in 2014, 66% of Americans were found to be confident and very confident in the safety of the US food supply, down from 70% in 2013 and 78% in 2012 (IFICF, 2014). Similarly, an earlier US study in 2011 reported a similar trend, as 73% of Americans were "more concerned today than they were 5 years ago about the food they eat," up from 65% in 2010.

Outcome scores for similar 5-year changes are provided in Table 7.18. Ideally, the surveys ask the same questions at the same time for the same period, which is the case for the European countries where data were available. Superior scores were given if two-thirds (66%) of respondents said that the food was safer. Responses below 40% received a poor scoring. The United States, Australia, France, and Italy performed poorly, while Canada and Ireland had the highest performances.

Table 7.18 Change in public trust in food safety over a 5-year period.

Country	Share of public who believe food is safer today (%)	Score
Australia	28	Poor
Austria	52	Average
Belgium	56	Average
Canada	67	Superior
Denmark	49	Average
Finland	57	Average
France	33	Poor
Germany	44	Average
Ireland	66	Superior
Italy	35	Poor
Japan	No survey found	
Netherlands	51	Average
Norway	No survey found	
Sweden	52	Average
Switzerland	No survey found	
United Kingdom	59	Average
United States	27	Poor

Sources: Corporate Research Associates Inc. (2012); FSANZ (2008).

National Food Safety System Performances Compared

This world ranking study ranks food safety performance by tier. The top six countries earned Tier 1 grades owing to their comparatively progressive (responsive) and superior (outcome) scores. The next five countries subsequently received Tier 2 grades, and the bottom six nations Tier 3 grades. The overall world ranking of food safety performance, led by Canada and Ireland and then France, is illustrated in Table 7.19. (See Table 7.20 for detailed scores across all 10 food safety performance metrics.)

Table 7.19 World ranking of food safety performance.

Country	Mean	Tier
Canada	2.6	1
Ireland	2.6	1
France	2.4	1
United Kingdom	2.33	1
Norway	2.33	1
United States	2.3	1
Japan	2.22	2
Netherlands	2.2	2
Finland	2.2	2
Denmark	2.2	2
Austria	2.2	2
Switzerland	2.11	3
Sweden	2.1	3
Australia	2.1	3
Germany	2.1	3
Italy	2	3
Belgium	2	3

Source: Adapted from The Conference Board of Canada.

What This New 2014 Version Means

Based on the 10 food safety performance metric benchmarks assessed in this study, Tier 1 countries performed very well compared with their international peers. Canada in particular earned excellent grades in most food safety performance metrics, though work remains to improve its performance by reporting on chemical risks in food consumption (TDS), more frequent nutrition and dietary studies, and additional improvements to traceability and radionuclide standards.

World food safety ranking methodologies, including this one, would benefit from additional primary data and objective indicators. This would also improve their effectiveness and reduce potential controversy surrounding such rankings. In particular, any subsequent global ranking study should consider the development of survey instruments to gather adequate comparable national evidence (and consensus) on food safety, such as on the implementation of food safety policies, governance, education and training, and surveillance, among others.

Table 7.20 World ranking of national food safety performances by metric.

Country	Pesticide use	TDS*	FBI*	Consumption	Capacity	Recalls	Traceability	Radionuclides	Allergen	Trust	Result (mean)	Ranking (tier)
Australia	Progressive	Moderate	Average	Poor	Superior	Regressive	Average	Progressive	Progressive	Poor	2.1	3
Austria	Progressive	Regressive	Superior	Superior	Poor	Regressive	Superior	Progressive	Moderate	Average	2.2	2
Belgium	Regressive	Moderate	Average	Average	Average	Regressive	Superior	Progressive	Moderate	Average	2	3
Canada	Progressive	Moderate	Superior	Average	Superior	Progressive	Average	Moderate	Progressive	Superior	2.6	1
Denmark	Progressive	Regressive	Poor	Average	Superior	Moderate	Superior	Progressive	Moderate	Average	2.2	2
Finland	Progressive	Moderate	Poor	Superior	Average	Regressive	Superior	Progressive	Moderate	Average	2.2	2
France	Moderate	Progressive	Superior	Superior	Superior	Regressive	Superior	Progressive	Moderate	Poor	2.4	1
Germany	Progressive	Regressive	Poor	Poor	Superior	Moderate	Superior	Progressive	Moderate	Average	2.1	3
Ireland	Progressive	Moderate	Superior	Poor	Superior	Progressive	Superior	Progressive	Moderate	Superior	2.6	1
Italy	Regressive	Progressive	Poor	Average		Progressive	Superior	Progressive	Moderate	Poor	2	3
Japan	Regressive	Moderate	Superior	Superior	Superior	Regressive	Average	Progressive	Progressive		2.22	2
Netherlands	Regressive	Moderate	Average	Superior	Superior	Regressive	Superior	Progressive	Moderate	Average	2.2	2
Norway	Progressive	Regressive	Average	Average	Superior	Moderate	Superior	Progressive	Moderate		2.33	1
Sweden	Progressive	Moderate	Poor	Average	Average	Regressive	Superior	Progressive	Moderate	Average	2.1	3
Switzerland	Moderate	Regressive	Average	Poor	Superior	Regressive	Superior	Progressive	Progressive		2.11	3
United Kingdom	Moderate	Moderate	Superior	Superior		Regressive	Superior	Progressive	Moderate	Average	2.33	1
United States	Progressive	Moderate	Superior	Superior	Superior	Moderate	Average	Regressive	Progressive	Poor	2.3	1

* Insufficient data.

It is important to recognize that none of the countries included in the survey had a performance that would suggest neglect by food regulators or the food industry. Italy and Belgium, countries with the lowest overall scores, still have very high food safety standards relative to the rest of the world. Benchmarking in food safety involves comparisons among countries that emphasize key patterns of similarity and difference. Results of this study should be considered in relative terms. Indeed, the comparative analysis itself among countries included in the sample should be of value to consumers, industry, and food regulators alike.

This 2014 benchmarking assessment follows similar world ranking studies from 2008 to 2010. Aside from the incidences of reported illness by food-borne pathogens and the rate of use of agricultural chemicals, results are not comparable with these previous rankings as the indicators and methods have changed.

Unlike 2008 and 2010, the choice of metrics in this survey may have contributed to an increase in variances among European-based countries. Typically, because of the EU's continental approach to risk mitigation and intelligence, differences in food safety performances between European countries are more difficult to detect. While France, Ireland, and the United Kingdom performed relatively well in this survey, Germany's and Belgium's performances are lower ranked.

Canada's performance is impressive. Since 2008, Canada's performance has always been in the top tier. Results in this survey are consistent with the 2008 and 2010 reports, even if different metrics were captured. Island countries like Japan, Australia, and the United Kingdom have historically been good performers over the years. This year though, the report suggests that both Japan and Australia's performances have dropped from 2008.

Most federal food regulators included in this survey have been affected, in one way or another, by the most recent global recession. Due to fiscal restraints, risk surveillance capacities have likely affected many countries since the two initial surveys. For more than a decade, budgets to support food safety policies and procedures have increased significantly. The global recession pressured most countries to revaluate how consumers were protected from potential food safety risks. To establish a strong correlation between budgetary readjustments and any food regulators' capacity to mitigate and communicate risks remains a challenge. Again, the current survey did not look at how governments were investing in food safety systems. It looked only at output and externalities.

Since the initial 2008 report, access to data has improved substantially. Given how important food safety accountability will become for industry, future surveys may include data from industry to assess involvement in food safety systems. The recalibration of the public sector around the world may have compelled some countries to seek not only more sustainable and long-term effective options but also more affordable options like self-regulation and/or self-reporting. Even if such an approach remains controversial, many food regulators anticipate that industry will play a larger role in making companies more accountable to themselves. The difference between self-regulation and accountability is significant. Results of this survey suggest that the role of the state will likely remain prominent around the world, but industry may play a larger role in scoping future risk intelligence efforts. The horsemeat recall in Europe in 2013, for instance, has made a clear case for an increase in supply chain-based scrutiny. This can only be achieved by making industry more accountable to itself without succumbing to a self-regulatory regime. Such an outcome is less than desirable, at least for countries included in this survey.

In essence, this report points to areas where countries can improve their performance. Most notably, as global food safety systems mature over time, benchmarking reports will increasingly reward consistency in high performance and transparency. To improve and support future food safety benchmarking efforts, funding harmonized food safety data collection is recommended, as is hosting a food safety summit for nations to find consensus on common robust food safety performance measurements, drawing on metrics from this study, among others.

8

The Future of Global Food Safety Systems and Risk Intelligence

The *Food Safety Performance World Ranking Initiative* project was, from its inaugural year, designed to report on what *was* not what *could* or *would be*. After all, a forward-looking analysis of a food safety regulator's performance would be ridiculous; no one can predict all food safety crisis nor can untested regulations be judged until they are tested. It is not to suggest that the *Food Safety Performance World Ranking Initiative* is strictly backward looking, since food safety laws change relatively slowly (Produce Safety Project, 2010).

While the *Food Safety Performance World Ranking Initiative* compares, grades, and ranks each of the 17 countries based on how they have performed, the discussion in the previous six chapters has been, to an extent, forward looking. This chapter is entirely forward looking.

Given the enormous changes that have occurred in the agricultural sector in the last 60 or so years and the enormous changes that have been forced upon modern food safety systems—a regulatory development not much more than 150 years old (World Health Organization (WHO) and Food and Agriculture Organization of the United Nations (FAO), 2006)—it is safe to say that the future of global food safety systems will be incredibly different in as few as 40 years. While making such far-reaching predictions would surely entertain readers in half of that time, this chapter takes the safer course and only points to a few important trends for the next few years.

Notable among these trends are the ever-evolving agricultural production strategies. In light of the seemingly increasing pace of change in the logistical, organizational, and economical forces that shape the modern agricultural industry, two (opposite) examples of production strategies are explored in some details. This discussion is bookended by a brief overview of the managerial and organizational forces that will shape and develop from these distinct methods of production.

To draw further distinctions between the "traditional" and "modern" forces of agricultural production (factory farming and *Terroir*), this

Food Safety, Risk Intelligence and Benchmarking, First Edition. Sylvain Charlebois.

chapter also discusses the potential, worry, and current reality of genetically modified organisms (GMOs). While the modification of crops (or animals) is hardly new, the subject of GMOs invites enormous debate worldwide. They ask massive questions about food safety not only from a scientific but a regulatory point of view. Moreover, current GMO crops are the subject of intense legal and economic battles fought in courts and regulatory tribunals. Finally, there are moral and philosophical questions that need to be answered as the technological ability grows and the potential (for good and bad) of GMOs continues to expand.

Finally, this chapter ends with a review of five (but discussed as four) current and systemic risks to the current food safety system. These five (increased mobility and system complexity, rising scale of nonproducer urbanization, increased pace of risk changes, new divisions of responsibility between private sector and government players, and the increased risk aversion of the public) are both related and separate from the other discussions.

What is most important to recognize from these systemic risks and the dangers that they currently present is, in fact, the shifting global food safety system. In the modern, globalized world of big-business agrifood and technological promise, food safety systems must continue to adapt in order to safeguard the consumer. While elements of the *Food Safety Performance World Ranking Initiative* point to current international best practices and areas for improvement, this chapter should make clear that the role of the public sector food safety regulator is changing. In response, every consumer should realize that; so too should the role of the modern food consumer. Ultimately, it will be up to the consumer and citizen to ensure their continued access to safe, nutritious, and sustainable food and agricultural industries.

Changing Agricultural Production Strategies

Since the 1950s, there have been tremendous changes in how agricultural producers grow, market, and sell their products to primary and secondary consumers. This evolution affects small family farms, large-scale industrial producers, corporate agricultural players, and whole industries. Since the turn of the new century, there have been enormous advances in information and biological technology, globalization-enforced trade liberalization, and shifting consumer demands for environmentally sensitive, healthy, nutritious, and safe food. In response to these ongoing changes, the shift in agricultural food production management in the 20 years between 2000 and 2020 could overshadow the enormous changes of the past 60 years since the 1950s.

At the turn of the twenty-first century, predictions for the shape of the food sector in the next 20 years focused on the division between agricultural commodities and distinguishable agricultural products. In 1999, Gary Brester and J.B. Penn predicted that:

> Over the next 20 years, farms and ranches will gravitate toward one of two production structures. The first type of production structure will be similar to many current farms and ranches in the undifferentiated commodity products will continue to be produced. Only low-cost producers will survive in this sector. A second category of producer will also evolve. Farms in this category will produce differentiated, identity-preserved products that focus on certain product attributes and consumer demands.

In 2010—half-way through Brester and Penn's 20-year prediction—their hypothesis has come true. Large-scale, low-cost corporate farms represent ever-increasing amounts of production from the agricultural sector. On the other hand, consumer-driven social movements and efforts by traditional farming communities and small-scale producers have opened up boutique industries on the other extreme. Two Canadian examples—One Earth Farms and Saskatchewan's *Terroir* opportunity—demonstrate these twenty-first-century agricultural strategies.

One Earth Farms

One of the most unique projects in the Canadian agriculture sector was brokered between two unusual agriculture partners to Canada's West. In March 2009, Toronto-based investment fund and wealth-management company Sprott Inc. announced the creation of One Earth Farms Corp. In partnership with First Nation bands in Western Canada, One Earth Farms plans to create a massive corporate farm using "world-class" farmland located in areas set aside as Indian reserves. Sprott Resource Corp. (a publicly traded Sprott Inc.-managed natural resource company) started One Earth Farms, kick-starting the new agribusiness with $27.5 million (Canadian) (Bambrough, 2009).

There are two especially engaging components in this project. First, the planned creation of a 50 000-acre farming operation within 1 year, growing to "the largest, most efficient, operating farm in Canada" (Bambrough, 2009) is bound to turn some heads. Even if such remarks are best taken with a grain of salt—the exuberance of a just-launched investment project or in order to account for lower-than-expected uptake or expansion—One Earth Farms could become a significant stand-alone player in the agricultural sectors that it enters. The second and more important issue concerns One Earth Farms' business model. In explaining the motivation toward investing $27.5 million (Canadian), Sprott Resource management:

> [Believed] that the timing for this venture is opportune. Global trends continue to impact food supplies, as arable land continues to decline, fresh water remains in short supply and various regions of the world are experiencing severe, recurring droughts. In addition, the global credit crisis has impacted the financing available to farmers and will negatively impact crop production in the short term. These factors, combined with a global population that continues to rise, are creating food security issues and in turn fuelling substantial farming investment demand globally. (Bambrough, 2009)

Once again—even parsing the language with a recognition that Sprott Resource is both announcing and justifying a multimillion-dollar investment—Sprott Resource clearly believed that agricultural commodity prices will continue to rise, that good farmland is valuable, and that this partnership deal represents a business model, which the company believes will make money.

Is (or, was) Sprott Resource correct? Likely, that question is best answered by the Toronto Stock Exchange, Sprott Resource shareholders, and the company's management and partnering First Nation bands. After all, in One Earth Farms' 2009 fiscal year, it had a gross loss of $419 000 (Canadian) (Sprott Resource Corp., 2010a). In the first three fiscal quarters of 2010 ending on September 30, 2010, One Earth Farms suffered a further gross loss of $10 000 (Canadian), although it had a gross profit of $10 000 (Canadian) during the most recent 3-month quarter (Sprott Resource Corp., 2010b). Nevertheless, setting aside the short-term economics of developing a massive farming operation, between March, 2009 and November, 2010, Sprott Resource's $27.5 million (Canadian) has netted it a 66.7% (80% fully diluted) interest in One Earth Farms, which now has 2.1 million acres of cropland in Alberta, Saskatchewan, and Manitoba. Ultimately, this level of agricultural potential—especially considering the partnership with First Nations bands—makes One Earth Farms worth watching.

Despite a pre- and post-contact history of agricultural trade and farming, Canadian First Nations bands started the twentieth century with several constraints. In Canada's Prairie region, First Nations signed reserve treaties in a state of economic and social difficulty as a result of the previously relied-upon bison's near extinction. While a tradition of farming had existed: "When they settled on their reserves, the Indians urgently needed to reacquire the skills needed for fixed market agriculture.

Indians knew the land, climate and soils intimately, but not from the perspective of commercial grain farming." Furthermore, the Canada *Indian Act* denied usual property rights to reserves, and the benefits usually ascribed to ownership and control of a (potential resource). Finally, First Nations bands did not have and were not provided with access to the capital necessary to buy needed tools (nor were they given the tools themselves) in order to build prosperous or sustainable reserve farms (Ward, 2009). In essence, the bands were stuck in a Catch-22 situation:

> Land was relatively plentiful, but with the switch to intensive agriculture, and given the immature technology, physical and financial capital became more important. Prairie Indians fared badly insofar as access to capital was concerned. The only way in which they could have enhanced the small treaty payments was through retained earnings, which never became sufficient for growth. (Ward, 2009)

So, in addition to a dramatic change to their natural resources (bison) and the loss of their social and economic foundation, First Nations bands had less expertise than their European extraction neighbors on running a modern farm, less access to capital and other necessary tools, and were operating under paternalistic, racially targeted legislation that denied them the rights to create, invest, and achieve generally understood capitalist goals.

Compared with early twentieth century realities, the potential benefits of partnerships between corporate agriculture experts and leased First Nations band land (with "job training programs for First Nations persons") (Bambrough, 2009) to unlock agricultural potential is an exciting development in the agricultural industry. Clearly, this model—while potentially beneficial to individual status Indians and their communities—is not a replacement for decades of racial inequality and generally poor agricultural development on reserves. Instead, One Earth Resources is a new form of partnership attempting to capitalize on otherwise underutilized land to build a large-scale, corporate-style operation. The wholesale adoption of factory farming operations represents, perhaps, a uniquely Canadian model of developing the next generation of low-cost super farms.

***Terroir* Potential**

If increased commodification, large-scale corporate farming, and a race toward lowering costs represent one possibility, there is another opposite model. In Europe and the Mediterranean, agricultural communities use the philosophy of *Terroir* to connect their products, culture, and local characteristics with the agricultural products that they take to market. The idea—the opposite of creating generic, lowest-cost-wins products—is for communities to take ownership of local specialities and to package their culture, history, and method of production inside of a community brand, which can then be attached to a now-distinct product. For example, consider Parma, Italy; Vidalia, Ga.; or Bresse, France. These communities are better known as brand names for the ham, onions, and chickens that these communities produce. One need only reflect on Champagne, France—one of the penultimate *Terroir* products—to see the power of this idea.

While there is certainly a romantic view that safeguarding the traditional lifestyle, culture, and production methods of ancient agricultural communities is good in and of itself, *Terroir* also offers enormous economic potential and is having a growing impact on world food trends. In a modern, globalized world, with widespread commodification of agricultural products and worldwide product markets, the evolution of our global versus local instincts presents an opportunity, in particular, for smaller-scale or community-based producers. If these players market or adapt their production and processing practices to reflect consumption dynamics, they can take advantage of opportunities created by these global–local interactions (Wilhelmina et al., 2010). It seems self-evident that trends toward gastro-tourism, buying foods locally, increasing traceability (for food safety, general information, and branding), and increasing menu depth are all based on a search for nutritional authenticity. If this is the case, the *Terroir* philosophy is ideally suited to fill this (growing) need. After all, *Terroir* is based on offering specialty, local, ethnic food products to respond to consumer demands (Blay-Palmer and Donald, 2006). Local foods, or *Terroir* products, are often conceptualized as authentic products that symbolize the place and culture of the destination (Sims, 2009). Consumers also have greater consumer confidence in the quality of regionally produced foods (Bruwer and Johnson, 2010). Furthermore, consumers often identify health, environment, and support for local farmers as their primary motivators for local food product consumption (Hamzaoui Essoussi and Zahaf, 2009), characteristics also inherent to popular conceptions of *Terroir* products. Most importantly, more and more consumers want to consume food that has a story, and they are buying it when they find it (Campbell, 2009).

Terroir represents the antithesis of large-scale, corporate, and low-price commodification of agriculture. While certain products seem naturally more disposed to *Terroir*-like branding (meat, cheese, wine, and produce) than others (e.g., the former seem inherently less like commodities than wheat, oats, corn, and barley), *Terroir* offers the potential for diversification or the active pursuit of a different market. For a Canadian example, consider Saskatchewan: Currently, raw, unfinished products dominate the province's exports. However, there is an existing, increasing, and largely unmet demand for value-added products in some of Saskatchewan's largest export markets. According to Agriculture and Agri-Food Canada, agriculture exports to some of Saskatchewan's largest markets are trending in favor of processed and value-added products (Agriculture and Agri-Food Canada, 2010). However, between 2007 and 2009 the province only exported an annual average of $850.3 million (Canadian) in processed products, accounting for an average of only 10.6% of total agrifood exports (Saskatchewan Ministry of Agriculture, 2010). Clearly, Saskatchewan faces an obvious opportunity to diversify its products and agricultural exports, while, at the same time, pursuing a very different agricultural model than Sprott Resource's One Earth Farms.

Possible Effects on Global Food Safety Systems

In an era of change, global food safety systems will have to react and adapt to changing production methods and the evolving needs and demands of consumers and producers. Naturally, this evolution will have to occur in the context of the real changes and needs, many of which are unknowable and difficult to spot in advance.

While hindsight may be 20/20, certain issues are certain to arise. Using a simple comparison between massive, commodity-type agribusiness, and small-scale *Terroir* models, food safety agencies will certainly have to deliver services across a rapidly polarizing industry.

Terroir producers will require services closer to the farm. By their nature, small-scale producers will require more inspection personnel, simply because there will be many small operations, versus fewer larger ones. The task of managing these needs in an era of tight budgets suggest that regulatory and technological and advances can only support so much innovation: At the end of the day, more inspectors may be the solution. While the task of inspecting these operations may be labor intensive, it is not necessarily one that will represent new systematic risks to the food safety system. Although the ideal amount of risk is always zero, the reality is that a *Terroir* producer will not produce enough goods to constitute a systematic risk to the public. In part, this is because of the nature of the good: *Terroir* is based on small-scale, niche operations, the opposite of large-scale agribusiness commodity producers. In this sense, with a low (systematic) risk and high personnel requirements per amount of goods, it might make sense for food regulators to create special regimes to take into account the unique needs and risks of *Terroir* producers, compared with large- or medium-scale operators—not to reduce regulations but to ensure that the regulations and oversight that exist make sense for the particular needs of consumers and the characteristics of the industry.

On the subject of regulation, *Terroir* producers may be interested in securing government regulation and protection for authentic *Terroir* products and production methods. While such policies are not unknown in Europe, they are new in Canada and the United States (US). Nevertheless, the recent proliferation of unregulated labels (the debate over what constitutes "organic" products in Canada, comes to mind), *Terroir* producers will want to protect their brands and products from non-*Terroir* imitators.

On the opposite end of the spectrum, the increasing proliferation of industrial agribusiness commodity producers suggests that increased regulatory scrutiny is also necessary. Like *Terroir* producers, these corporate interests will need specific, needs- and consumer safety-based regulations that address the specific characteristics of the industry. While it is certainly in the interest of large corporate producers to adhere to the highest food safety standards, sadly mistakes, oversights, and unforeseeable events mean that food safety crises can, have, and will occur. As the industry becomes increasingly commodity based, industries of scale will continue to develop. Given the near-certainty that food safety concerns will arise, the systematic risk of these changes cannot be overstated. Accordingly, not only food safety regulators require the necessary resources to monitor and react to these issues. While the development of these industries will dictate what specific resources are necessary, the ability to institute and execute rapid recalls, tracing the programs, and communications to the public are the minimum tools for these agencies.

While the industry appears to be shifting toward low-cost, commodity-based producers on one end and ultralocal, authentic *Terroir* products on the other, the important reaction by food safety agencies is to recognize this shift and to move to proactively address new concerns and to regulate the industry in a smart, consumer-focused way. That the examples here may require different models, the important element is to maintain the public's access to safe, healthy foods and to communicate effectively to keep the public informed.

Other Policy and Managerial Implications

Between the One Earth Farms example and the potential for Saskatchewan to create a *Terroir*-based industry, the common trend is greater specialization in the low-cost, high-volume commodities model of the small-scale, niche-market products. Nevertheless, both examples share the common foundation of shifting organizing principles. One Earth Farms created a three-province-spanning megafarm (partnering with local First Nations bands). The model for Saskatchewan *Terroir*, on the other hand, could be built around a cultural- or community-based organization in partnership with the community, cultural, and public stakeholders (Charlebois and MacKay, 2010). This organizational system follows the Brester and Penn prediction that "[organizations will] change their primary objective from lobbying for traditional commodity programs to providing resources and services needed by producers to cope with change and to expand profit opportunities." The One Earth Farms model also fits this prediction, if one considers the partnership aspect with First Nations bands that use One Earth Farms' resources, expertise, and one-desk point of sale.

Ultimately, as globalization reduces the logistical barriers between individual producers and worldwide consumers, the agricultural sector is becoming more efficient, more competitive, and less forgiving of managerial or policy roadblocks. Michael Porter laid out an organizational theory underpinning this type of strategic business management as early as 1980. In his *Competitive Strategy*, Porter broke successful companies into three categories: low cost, differentiation, or focus. Companies that employ the first strategy do so by achieving a competitive advantage on cost, usually by creating economies of scale and new technology or the outsourcing or contracting of jobs to low-cost bidders. One Earth Farms is an excellent example of this first strategy, while Saskatchewan *Terroir* is a little more complicated to classify. The second and third strategies are, respectively, differentiation and focus. Differentiation occurs when "a producer incorporates features into goods or services that cause buyers to prefer that firm's product/service over those of others." This is a good definition of *Terroir*; however, the local product philosophy also encompasses elements of the focus strategy. Focus strategies target niche markets through particular product characteristics. For example, the "Certified Organic" label allows organic apple farmers to target potential apple buyers that are looking for pesticide-free produce. *Terroir* products bridge the both of these strategies, as *Terroir* producers both imbue their products with special, customer-targeted characteristics, while, at the same time, targeting niche—or, at least, noncommodity—markets that attract a particular type of customer. Based on these three business strategies, Brester and Penn argue that either low-cost or value-added operators, with few surviving businesses models in between, will increasingly dominate the agribusiness sector.

At this point in the twenty-first century, forecasting the continued growth and efficiency of the agricultural sector is not a daring prediction. Agricultural production strategies will continue to evolve, especially given the expectation that the forces of globalization will continue to reduce the logistical—if not actual—distance between producer and consumer and the continued growth of large-scale agribusiness based on product commoditization. From the point of view of the commodity producer, being the lowest-cost seller ensures the best opportunity to increase market share. However, these types of producers (assuming that megafarms do not turn into monopolies) will remain price takers based on global availability and demand. On the other hand, if Brester and Penn are correct and *Terroir*-type value-added products represent a viable alternative, these producers may be freed—by virtue of their unique and not readily imitable desirability—to become price makers. While *Terroir*-type products may be an attractive (if not historically romantic) future to imagine, these types of developments carry their own risks. Brester and Penn point out that, "Strategic business management abilities will be especially critical for farms that gravitate toward identity-preserved products." Indeed, when your product is your market maker, the margin of error shrinks. Consider the difficulty for small farmers when they must, effectively, start running a marketing as well as an agricultural small business. In addition to keeping up with general marketing trends to ensure the desirability of their unique, *Terroir* products, these producers also bear the risks of running (comparably) small-scale farming operations. Moreover, in addition to the difficulties of running a *Terroir*-type farm, one must still produce the products in keeping with the general philosophy. For some this means preserving important cultural, production, and quality traditions. On the other hand, this also means doing without some of the potential benefits from the onward march of technological progress.

On the Question of GMOs

In addition to a discussion of the possible business strategies (low cost, differentiation, and focus) that results in two increasingly divergent agricultural product types and businesses (low cost and *Terroir*-type value added), Brester and Penn note the increasing proliferation of widespread adoption of genetically modified crops. This shift in agricultural production is a by-product of as well as a driver toward globalization and the increasing movement of agricultural products worldwide. Eventually, "Crops and livestock products may be genetically engineered to provide animals and humans with needed vaccines and pharmaceuticals"—a situation whereby a GMO apple a day may really keep the doctor away. While "biotechnology now allows the production of crops that have specific attributes sought by consumers," some view GMOs in a negative light. In fact, Brester and Penn attribute the rise of GMO foods as a creator of "new market opportunities ... for food and fibre processors that require identity preservation of crop and livestock products with specific attributes." The opportunities promised by GMOs as well as the backlash that this technology has created will continue to represent a powerful pressure point in the current and future global food safety system.

The *Food Safety Performance World Ranking Initiative* has never considered GMOs as part of its food safety criteria or analysis. They were originally excluded from consideration due to the absence of conclusive evidence that agricultural products that comprise or include GMOs posed a health threat to consumers. As a result, when the methodological framework for the most recent 2010 *Food Safety Performance World Ranking Initiative* project was constructed, GMOs were again excluded in favor of the original set of criteria. Because the same criteria are studied, evaluated, and analyzed in each iteration of the *Food Safety Performance World Ranking Initiative*, each country's section-by-section grades can be directly compared and analyzed in relation to the previous results. Furthermore, by not changing the section-specific criteria, the final world ranking is also comparable across reported years because each country is comparatively graded against its peers across the same areas year by year. However, despite the exclusion of GMOs from the food safety criteria in the 2010 report and its predecessors, the GMO debate is a real and growing part of the global food safety discussion. There are two large issues motivating the debate: trade and capability.

In brief, part of the debate over GMOs is a result of the globalized, transnational agricultural industry. While products can and do move across international borders with increasing frequency, sovereign nation states (including the European Union (EU)) retain the exclusive power to regulate food safety. Despite CODEX Alimentarius standards, World Trade Organization (WTO) rules, and United Nations (UN) treaties (including the powers of the World Health Organization and the Food and Agriculture Organization of the United Nations [FAO]), national boundaries can quickly become trade barriers if a state so chooses. Therefore, a country that wishes not to grow GMOs internally, based on food safety risks, for example, might feel threatened by the import of GMO products from foreign countries. In effect, where GMOs are concerned, some countries worry not about whether a *specific* imported agricultural product is safe but whether the *entire* imported agricultural food sector is safe. Naturally, a GMO-wary country can be expected to place import barriers on GMO products for internal consumption. With the ever-increasing proliferation of the international agricultural marketplace, the concern is that GMO products can easily spill over from a GMO-accepting country into a GMO-wary country. While important to keep in mind, the GMO trade concern is less of a cause of the debate over GMO use and more of a by-product. The principle debate is not whether GMOs are possible—after all, they clearly exist—but whether GMOs should exist.

The question of genetics in agriculture is not new. In fact, humans have been creating new and hybrid organisms since the dawn of agriculture—sometimes deliberately but often enough by accident. However, the birth of gene manipulation as a science has invigorated the debate. On the innovation front, genomics has played a significant part in augmenting the world's capacity to grow foods. Norman Borlaug, whose Mexico experiments resulted in the "semidwarf" strain of wheat, is widely credited with quadrupling India's wheat (and later, rice) production and bringing the subcontinent, China, and other parts of the world out of constant, drought-induced famine. Since the 1960s, enormous technological and biological progress has rapidly occurred. So, while genetically engineered crop seeds have existed for years, agricultural is now at a transition point where digital codes will soon be replaced by life codes (Moody, 2004). In effect, as important as the information technology revolution has been to the industry, producers, and consumers, the potential offered by genetic manipulation could be the next big evolutionary step.

However, the use of the word "could" is important: Embracing biotechnologies can be a double-edged sword. Previously, the approach toward agriculture was a linear thought process involving three "F"s: food, feed, and fiber. However, methodologies such as genomics can change the relationships among these three elements, especially when combined with other radical shifts in the sector's other theoretical models. Furthermore, the science of GMOs may or may not, in practice, increase the risks to which consumers are exposed. Either way, it is certain to alter those risks in many ways. Finally, and possibly most importantly, the spread of GMOs is certain to alter the ways in which consumers perceive products crafted by new technologies. In a 2003 report, the Organisation for Economic Co-operation and Development (OECD) outlined this twofold risk from a scientific and health-based point of view:

> Biotechnology—which is based on modification of living material—raises a number of new questions linked to potential interactions between species inside an ecosystem. Genetically modified foods also raise a number of health risk assessment issues, notably concerning the effects of antibiotic-resistant genes and the introduction of unexpected alterations in nutrients. Assessment of such potential interactions is expected to become even more complex with the next generation of GM foods, where new traits will often

> be created by inserting multiple genes, making any reference to "traditional" counterpart products more difficult. (Organisation for Economic Co-operation and Development, 2003)

However, the health issues raised by the OECD also raise philosophical concerns. Likely, most consumers have—at one time or another—worked in a garden, planted a tree, or potted some flowers. The relationship between a seed—natural and timeless—and a plant is something that everyone should be able to grasp. If one feels that the genetic modification of an agricultural product somehow affects the philosophical sense of what a seed of agricultural product is or that the manipulation somehow makes the resulting product unsafe for consumption, the seed-to-plant relationship that consumers have now will change or fade. In essence, between the philosophical (should we) and the safety (can we) questions, GMOs will play important and ever-growing roles in the future of agriculture and food safety decisions.

In Canada, the federal government, through Health Canada, regulates and assesses the safety of GMO and other "novel" foods (Health Canada, 2010). Generally, "Health Canada has taken the position that [GMO] foods are just as safe as conventional foods," although producers are required to obtain approval before selling these products (Health Canada, 2010). This is not to suggest that GMO foods have not stirred some controversy in Canada. Anti-GMO groups like Greenpeace and the Council of Canadians have argued that GMO products may represent a health risk, in part because they do not have long-term health studies on the health impacts of including GMO foods in one's diet. However, in addition to questions about the long-term health impacts (for better or worse) or GMOs, their widespread introduction is also having important impacts on agribusiness. In the precedent setting, *Monsanto Canada Inc. v. Schmeiser* case, the Supreme Court of Canada confirmed intellectual property rights for GMOs. As a result, the court has set a significant precedence in the evolving agribusiness models that are possible with patentable seeds.

The facts of *Monsanto Canada Inc. v. Schmeiser* are fairly straightforward. In 1998, tests of Percy Schmeiser's Saskatchewan canola fields found that his crop comprised between 95 and 98% of Roundup Ready Canola-brand canola, a GMO product that includes Monsanto Canada Inc.-patented genes. These genes are designed to make the canola seeds resistant to Monsanto's Roundup herbicide, which can then be used to control weeds and also eliminates the need for tillage or the use of other herbicides. Farmers who want to buy Roundup Ready Canola from Monsanto sign a licensing agreement, attend a "Growers Enrollment Meeting" and pay a $15-per-acre (Canadian) (in 1998) licensing fee. After the 1998 tests of Mr. Schmeiser's crop, Monsanto sued Mr. Schmeiser for patent infringement, because—Monsanto alleged—he had "never purchased Roundup Ready canola nor obtained a licence to plant it." While the original source of the Roundup Ready Canola seeds on Mr. Schmeiser's land may have been the wind (five neighboring farmers were using the product), the original trial judge decided that none of the suggested explanations offered by Mr. Schmeiser "could reasonably explain the concentration or extent of Roundup Ready canola [in his fields]." In a five-to-four decision, the Supreme Court found that Mr. Schmeiser had infringed on Monsanto's patent—which was itself valid because it covered the "genes and the modified cells that make up the plant" not the GMO plant itself—and that Mr. Schmeiser "was not an innocent bystander; rather, he actively cultivated Roundup Ready Canola" (*Monsanto Canada Inc. v. Schmeiser*, 2004).

Given that the Supreme Court upheld Monsanto's ability to patent elements (the genetically modified parts) of an otherwise generic product (canola), this decision will continue to have an important impact on the business models of the GMO-industry and producers who plant these products. Obviously, Monsanto's success at defending its patent for the GMO parts of Roundup Ready Canola provides an intellectual property rights framework that encourages other biotechnology companies to create new GMO strains. After all, if the courts found that Monsanto was unable to patent and control its products, while at the same time charging a premium for their benefits, farmers would have an incentive to cultivate their own non-Monsanto source for the seeds each year. (This, in effect, was what Mr. Schmeister was accused of doing.) However, since the court upheld Monsanto's rights to control its Roundup Ready Canola, Monsanto's business model of licensing agreements, per-acre fees, and specific sellers remains viable. Naturally, farmers can choose to plant whichever variety of, for example, canola that they wish. Nevertheless, if GMO products can provide enticing benefits to farmers now and corporate intellectual property rights are not changed, companies like Monsanto selling products like Roundup Ready Canola can be expected to brand and control basic farming inputs (like seeds) in greater numbers—the potential benefits of these products to farmers will be simply too great to ignore.

It is important to recognize the growing impact of GMO products in the agricultural sector. Currently, one of the issues that attracts the most attention is labeling standards. In Chapter 3, the evaluation of each country's labeling and the indication of allergens as part of the *Food Safety Performance World Ranking Initiative* formed an important part of the Consumer Affairs category. Since the *Food Safety Performance World Ranking Initiative* has not included GMO issues from a health point of view, GMO labels were not considered in that category. Nevertheless, there are distinctions between labeling approaches.

A 2004 estimate claimed that the number of different GMO products on Canadian grocery store shelves might total 30 000. The Canadian Food Inspection Agency (CFIA), which is responsible for packaging and labeling, has said that creating a labeling standard in Canada would be basically unenforceable because of the plethora of GMO products grown in Canada and the likelihood of GMO seeds mixing with non-GMO seeds in the field. As evidenced by *Monsanto Canada Inc. v. Schmeiser*, this includes most of Canada's canola crop. While the CFIA will test export-bound products for markets that require a label, none is currently required in Canada. In 2001, a private member's bill from Charles Caccia—a federal Member of Parliament representing downtown Toronto—that would have required GMO labeling was defeated in a free vote. The Canadian Federation of Agriculture has been against mandatory labeling requirements, concerned that "consumers will see the labels as a warning and avoid these foods." This may be a valid concern for the industry, regardless of whether the public's worry is well founded or not. According to a 2004 CBC News report, Greenpeace Canada, though not an impartial advocate, claimed that 95% of Canadians want mandatory GMO labeling. A 1999 Environics poll said the number was 80%.

In the United States, consumer concerns about the safety of GMOs are also an issue. In a 2010 survey commissioned by Deloitte LLP, only 21% of consumers are "not concerned" about eating GMO foods. This is a one-percentage-point decline from a 2008 survey. All told, while fewer consumers are "extremely concerned" (13% compared with 15% in 2008), over 24% of consumers are either "extremely concerned" or "very concerned" (Deloitte, 2010). This broadly mirrors the findings of the Pew Research Center's Global Attitudes Project which found only 37% of Americans supported GMOs when asked the following:

> Some people say that it is good to scientifically alter some fruits and vegetables because it increases crop yields to feed more people and is good for the environment. Others say it is bad to scientifically alter some fruits and vegetables because it could hurt human health and the environment. Which comes closer to your view? (Pew Research Center, 2003)

(By comparison, 31% of Canadians answered "good" (Pew Research Center, 2003).)

Anti-GMO movements in Europe have "exerted more pressure on their governments" and seem to have had more success. In response to the anti-GMO action, producers Nestlé United Kingdom and Unilever United Kingdom have removed GMO inputs from their products. The 2002 Pew Global Attitudes Project poll found that 65% of respondents in Great Britain, 74% in Italy, 81% in Germany, and an enormous 89% of respondents in France feel that: "Scientifically altered fruits and vegetables" are "bad" (Pew Research Center, 2003). Nevertheless, GMOs have been technically legal in the European Union since the European Parliament passed Regulation (EC) No. 1829/2003 in September 2003 (European Parliament, 2003a). However, the Act places greater restrictions on GMOs than either Canada or the United States. As a separate regulation, EC 1830/2003 mandates the traceability and labeling of GMO products (European Parliament, 2003b). In addition, as provided for in the "safeguard clause" of Directive 2001/18/EC—On the Deliberate Release into the Environment of Genetically Modified Organisms and Repealing Council Directive 90/220/EEC—"[a] Member State may provisionally restrict or prohibit the use and/or sale of [a] GMO" (European Parliament, 2001). As of November 2010, "Six member states currently apply safeguard clauses on GMO events: Austria, France, Greece, Hungary, Germany and Luxembourg."

If European anti-GMO advocates have had more success convincing politicians of the validity of their concerns than their Canadian or American counterparts, US lawyers have had more success (or, maybe, simply more cause) in winning GMO deregulation lawsuits than their Canadian neighbors. Similar to Canada's *Monsanto Canada Inc. v. Schmeiser*, the US courts have been increasingly considering GMO issues. However, unlike in Canada, these high-profile cases have dealt with food safety, not intellectual property. Monsanto Co. (the US-based global parent of Monsanto Canada Inc.) has faced two similar lawsuits dealing with GMO products, both originating in the San Francisco-based California North District Court.

Monsanto Co. v. Geertson Seed Farms dealt with the deregulation of Monsanto's Roundup Ready Alfalfa ("a variety of alfalfa that has been genetically engineered to tolerate the herbicide Roundup") by the United States Department of Agriculture's Animal and Plant Health Inspection Service. Geertson Seed Farms and environmental groups challenged that decision in court. The California Northern District Court agreed with the respondents, vacated the deregulation decision, forbid the Animal and Plant Health Inspection Service from rederegulating Roundup Ready Alfalfa before completing an environmental impact statement, and ordered a nationwide injunction against the planting of "*almost all*" Roundup Ready Alfalfa until the study was completed (*Monsanto Co. v. Geertson Seed Farms*, 2010). The United States Supreme Court overturned the scope of the district court judge's ruling because the decision:

> Abused its discretion in [preventing the Animal and Plant Health Inspection Service] from effecting a partial deregulation and in prohibiting the possibility of planting in accordance with the terms of such a deregulation. (*Monsanto Co. v. Geertson Seed Farms*, 2010)

However, the Supreme Court had decided a largely procedural question: Whether the District Court could enact nationwide bans on Roundup Ready Alfalfa and forbid the Animal and Plant Health Inspection Service from introducing a partial deregulation while conducting a (required) environmental impact assessment. Despite Monsanto winning the case (the Supreme Court agreed that the District Court overstepped its authority), the result of this decision is that "virtually no [Roundup Ready Alfalfa] can be grown or sold until such time as a new deregulation decision is in place" (*Monsanto Co. v. Geertson Seed Farms*, 2010). The Supreme Court's validation of the result—but not the scope—of the original ruling has opened the door to further court-enforced environmental study of the environmental and health impacts of GMOs.

Based partly on the Supreme Court's *Monsanto Co. v. Geertson Seed Farms* decision about GMO alfalfa, the California Northern District Court arrived at a slightly different ruling concerning Monsanto's Roundup Ready sugar beets. After a September 2009, decision in favor of an summary judgment that the Animal and Plant Health Inspection Service (APHIS) violated the *National Environmental Policy Act* by failing to conduct an environmental impact statement, in August 2010, the court vacated the decision to deregulate GMO sugar beets and sent the issue back to the regulator (*Center for Food Safety, et al. v. Thomas J. Vilsack, et al.*, 2010). However, the judge specifically criticized the government regulator, saying:

> APHIS's apparent position that it is merely a matter of time before they reinstate the same deregulation decision, or a modified version of this decision, and thus apparent perception that that conducting the requisite comprehensive review is a mere formality, causes some concern that Defendants [1] are not taking this process seriously. (*Center for Food Safety, et al. v. Thomas J. Vilsack, et al.*, 2010)

Among GMO-concerned advocates, this ruling was expected to tighten up a regulatory review on GMO deregulation in the United States (Gillam, 2010). However, as of summer 2010, GMOs represent over 95% of the US sugar beet crop (Pollack, 2010).

---

1 The *Center for Food Safety, et al. v. Thomas J. Vilsack, et al.* case was a lawsuit between the Center for Food Safety, the Organic Seed Alliance, the Sierra Club and High Mowing Organic Seeds (a seed company) for the plaintiffs, and Thomas Vilsack and Cindy Smith for the defendants. Mr. Schafer was sued in his official capacity as the Secretary of the United States Department of Agriculture, while Ms. Smith was sued in her official capacity as Administrator of the Animal and Plant Health Inspection Service (*Center for Food Safety, et al. v. Thomas J. Vilsack, et al.*, 2010).

The GMO question is certainly not going to be settled anytime soon. One must recognize the potential that is promised by GMO advocates could solve enormous agricultural problems in the present and into the future. According to the OECD, "Biotechnology has the power to improve human health, address environmental challenges, and change the way the world does business" (Organisation for Economic Co-Operation and Development, 2009). As *Monsanto Canada Inc. v. Schmeiser* demonstrated, the business opportunities are already well developed, as proprietary seeds open new business avenues for sellers. One of the largest corporate players, GMO-megacorporation Monsanto, describes itself as being about farmers:

> Billions of people depend upon what farmers do. And so will billions more. In the next few decades, farmers will have to grow as much food as they have in the past 10,000 years—combined. It is [Monsanto's] purpose to work alongside farmers to do exactly that.

Can Monsanto and other GMO producers fulfill these lofty promises and the potential that the OECD forecasts? Presumably, time and the continued study of GMO foods will eventually provide the answer. However, it is obvious that Monsanto and its peers have a vested business interest in ensuring the increased adoption of GMO-type products. Between consumer apprehension about GMO products, an absence of labeling standards (at least between Western democracies) and the extremely rapid expansion of GMO products on store shelves, questions should be asked about the future role of GMOs as they relate to food safety and the environment. We do know that these questions will also not be quickly answered. Nevertheless, the role of food safety authorities nationally and internationally will have to rise to the challenges presented, on one hand, by GMO producers promising twenty-first-century agriwonders, and, on the other hand, anti-GMO advocates worried about the impact of GMOs on the environment and consumer health. Questions about the diligence of the United States Department of Agriculture's Animal and Plant Health Inspection Service from *Monsanto Co. v. Geertson Seed Farms* and *Center for Food Safety, et al. v. Thomas J. Vilsack, et al.* raises concerns that the consumer's supposed representative and food safety guardian—government regulators—are not adequately studying GMOs in the United States. Furthermore, while scientific study on the environmental and health benefits of GMOs will doubtlessly prove critical to any widespread support among the public, to what extent should government regulation also adhere to nonscientific concerns among the general public? These philosophical questions are difficult to answer from a policy setting point of view: "While there seems to be agreement that the social process of risk handling needs to be 'open, transparent and inclusive' and should clearly acknowledge scientific uncertainties and take into account the validity of social concerns, there is no consensus on how this should be done in practice." Doubtlessly, in order to take advantage of any GMO potential and to minimize any potential hazards, food safety authorities must maintain a firm hand on existing and growing food safety risks.

Assessment of Current and Evolving Systemic Risks in Food Safety

The OECD's 2003 *Emerging Systemic Risk in the 21st Century: An Agenda for Action* report begins with a discussion of emerging systemic risks that include BSE, climate change, and GMOs among the five "few examples of important risk debates that have emerged in recent years." Whether because of changes in demographics, the environment, technology, or socioeconomic divisions, the OECD predicts that, "Our societies are ... increasingly subject to major risks" (Organisation for Economic Co-Operation and Development, 2003). As discussed in Chapter 5, it is important for countries to have risk management plans for analyzing and preventing risky food safety problems. However, the dangers present in food safety systems are not only evolving as issue-specific policies are implemented, but the types of dangers are also changing as well. The OECD highlighted five issues that will impact the future of (nonfood specific) risk management: increased mobility and system complexity, rising scale of nonproducer urbanization, increased pace of risk changes, new divisions of responsibility between private sector and government players, and the increased risk aversion of the public (Organisation for Economic Co-Operation and Development, 2003). These five issues are each excellent examples of new, possibly growing and certainly evolving systemic risks in food safety systems.

Increased Global Trade

Recognition of the increasing importance of international agricultural trade has existed throughout the discussions of food safety systems in this book generally and the *Food Safety Performance World Ranking Initiative* project specifically. However, the increased mobility of food products and the increasing international demand—for basic and luxury food imports—has created a globalized world where food safety risks have gone global while food regulators remain national.

In Chapter 6, discussions about international trade formed important parts of the discussion about the importance of traceability. As discussed, while traceability is not itself a by-product of the increase in the global food trade, the increasing prevalence of "food kilometers" between producer(s) and consumer has made traceability an important component of a twenty-first-century food safety system. Still, knowing from where food comes and to where it goes only helps solve problems after they have occurred—through product recalls or the like. Moreover, because of the increased technological, economic, and social shifts in modern agriculture, trade is helping to spread food safety risks globally.

Before the modern world of international trade, "geographical separation has been a key barrier to the spread of disease." In fact, thanks to modern agricultural transportation systems, "shipping tends to increase the amount of handling and the time between farm and consumer, which can give pathogens on the food more time to multiply, potentially amplifying the hazard for the consumer" (Buckley and Reid, 2010).

The concern that trade itself is dangerous is certainly shared by some critics of the CFIA. According to *The Globe and Mail* in November 2010, "Every day, a rising tide of foreign food makes its way onto Canadian grocery shelves, the vast majority of it entering this country untouched by federal inspectors." Moreover, "Critics say Canada's ability to safeguard its citizens from the risks of both domestic and imported food is falling behind," a charge that Rick Holley of the University of Manitoba seems to support. The Globe quoted Dr. Holley as characterizing food safety in Canada as "an accident" and one that the CFIA has "proven to be incapable of dealing with." In response to some of these criticisms and the concerns about its inspection record, the CFIA is improving its traceability regulations and systems. (These improvements are discussed in Chapter 6.)

Ultimately, increased traceability may be able to help food safety systems avoid the systemic risk posed by ever-increasing international trade. This fact is recognized in the CODEX Alimentarius standard on traceability (CODEX Alimentarius Commission, 2006). However, the concern is that increased traceability requirements could be a guise used to impose nontariff barriers to trade in order to protect domestic agricultural sectors from foreign competition (something what the CODEX standard specifically highlights (CODEX Alimentarius Commission, 2006)). An excellent example of the traceability as tariff argument is the ongoing dispute between the United States and the European Union over GMO products.

The United States, which had been a large exporter of corn to Europe before 1997, "largely stopped shipping bulk commodity corn to the EU because such shipments typically commingled corn from many farms, including genetically modified varieties not approved by the E.U." For example, what had been a 1.74 million ton market in Spain and Portugal—the two largest importers—had shriveled to below 174 000 tons in 1998–1999. By 2004, the European Union imported about 0.1% of the US's corn exports. The well-documented European distaste for GMO products might be explained by their experience with major food safety crisis, different cultural attitudes toward food, an absence of confidence in European food safety regulators, or the prevalence of widespread media and activist agitation by well-known groups. Alternatively, US actors including GMO producers, farmers, and politicians objected to the WTO that the EU's GMO rules were anti-free trade, without the scientific basis necessary for such barriers and ultimately protectionist. Ultimately, the EU's 2004 GMO-permitting rules failed to staunch US critics due to traceability and labeling requirements.

Is traceability a silver bullet for solving the systemic risks to food safety systems caused by increased international food trade? Probably not. However, it is doubtlessly a necessary tool to stem the crises that will surely emerge in the future.

Taking Everyone Off the Farm in an Era of Rapid Change

While the OECD considered the increase in urbanization and rapidness of risk changes separately, these two issues are linked to the systematic risks inherent to any specialization.

Consider first that in the next 40 or so years, the global agricultural system must expand its output dramatically. To begin, the world's population is expected to increase by about 2.3 billion people—about one-third. Add to that challenge the reality that fewer people are maintaining the traditional rural agricultural life whose labor force used to produce and consume food locally. Finally, there exists a host of other pressures with increases in biofuel needs, an increasingly developing-nation-focused sector and increases in demand for harder-to-produce protein as the developing world adopts a standard Western diet of products like beef (FAO, 2009). While the UN FAO forecasts that the potential for productivity and yield improvements by 2050 will be adequate to feed an increasingly hungry (and large) world, this will likely accompany an urbanization and nonlocal shift toward the agricultural sector:

> A vibrant agricultural sector has been the basis for a successful economic transformation in many of today's developed countries. … During these transformations, investment in agriculture created

> agricultural surpluses, kept real food prices low and helped stimulate overall economic growth. At the same time, overall economic development created new employment opportunities that helped absorb the rural labour surplus that emerged from the transformation of agriculture. Theoretically, the result is a transition from many, small subsistence producers to fewer and larger commercial farmers and a new equilibrium with fewer farmers, more non-farm employment and larger farm operations overall. (FAO, 2009)

However, with the increasing concentration and scale of the global agricultural sector, new risks as well as the level of danger that they provide rise.

The most obvious risk of the globalized food industry is the risk of terrorism. It is chilling to read forecasts published in the last decade by food safety experts, some of whom predict that the next 9/11 will occur through our food supplies (Macpherson and McConnell, 2007; Choe et al., 2009). Post-9/11, some countries have recognized the risks to agriculture presented by terrorism: Threats to food production, processing, and distribution systems can no longer be presumed accidental. Even though most food safety incidences are not human induced, some may be malicious. Agricultural systems have been considered as prime targets for terrorist groups aiming to reach a great number of citizens (Halweil, 2004). Because of the massive and panic-inducing potential of such an attack, it must surely be attractive. Additionally, this menace is particularly imaginable because of inadequate food safety architecture. While the events of September 11, 2001, certainly changed how we manage risks not only to public infrastructure and landmarks but also to our food production and distribution systems, the disaster changed how we perceive such risks.

In addition, while there exists a general risk of bioterrorism in global food systems, it manifests in different ways in different countries. North America may be more guarded against foreign risks because of the United States' preoccupation with bioterrorism and its own experience; however, in other countries, threats may originate from citizens as well. For example, in China, a food distributor poisoned its competitor's food products in the pursuit of increasing profits and local market positioning. This incident victimized many from across the country (Johnson, 2008) and was depicted as a terrible act of food sabotage (Anonymous, 2008). Still, the risk of explicit bioterrorism—malicious, intentional attacks against a country's food system—has increased. The 2007 World Health Report lists "the natural accidental and deliberate contamination of food" as "one of the major global public health threats in the twenty-first century" (World Health Organization, 2008). According to the United Kingdom's Centre for the Protection of National Infrastructure:

> The public and businesses within the food and drink sector now face a different threat—that of malicious attack, especially by ideologically motivated individuals and groups. This threat will manifest in a way which reflects the motivation and capability of these people. It will not follow the statistically random, and therefore predictable patterns of familiar 'hazards.' (United Kingdom Centre for the Protection of National Infrastructure, 2010)

Along with the rapidly familiarized threat of terrorism is the increased risk of price volatility in global commodity markets.

While hardly a new event, the rapid increase in commodity prices to produce price spikes that drive consumers out of the market has gained modern currency after the events of 2007 and early 2008. While there has always been price movements in perishable products and agricultural inputs are particularly difficult to guarantee (e.g., due to weather), the European Union believes that "[price] volatility has increased at least in some commodity markets." Moreover, there are numerous causes that suggest the modern, globalized world of the twenty-first century might be particularly at risk of further volatility. Climate change, world financial crisis (e.g., the credit crunch and corresponding recession) and financial speculation are all possible causes (European Commission, 2009). Probably more concerning are the limited remedies to such inflationary spikes in philosophically free-market economies. Even China—hardly an unregulated economy, even in 2010—has difficulty imposing price constraints on food: In reaction to inflation worries the government unveiled food subsidies, not price controls (McDonald, 2010). The risk of uncontrollable inflation in food prices represents a possible threat to the Chinese Communist Party's ongoing political control:

> The ruling Communist Party's ability to manage inflation could play a key role in the impending handover of power to a new generation of leaders, Mr. Broadfoot said. "They're going to be talking about inflation because it has a direct impact on social stability," he said. (McDonald, 2010)

The by-products of increasing change—even volatility—in a trade-friendly globalized agricultural sector is both a motivator for as well as a product of the move toward the increasing power of and need for large-scale agricultural producers. As illustrated by the difficulty if not impossibility of national governments protecting their citizens (and foreign consumers) against bioterrorism on the long term, risk profiles increase to the point where they are beyond any one agency's power to control. In recent years, Canadians have had faith that the government knows best when it comes to public health issues and rightly so. Our public health system has served us so well for decades that it has become unnatural to think that there is another, better model. However, while many Canadians believe we should rely solely on publicly funded authorities, the expanding scope of modern food systems is debunking such wishful thinking.

Shifting Food Safety Responsibilities Between Public and Private Sectors

In part because of difficulties inherent in their mandates, government regulators face increasingly difficult prospects in the near future. Whereas food used to be local by necessity—as a result of trade barriers or technological difficulties in shipping large quantities of perishable goods—international food trade is an enormous component of the agricultural sector.

> Last year, the [Canadian] import count included more than 33 million litres of apple juice from China; 11.8 million kilograms of pickles and relish from India and 4.9 million kilograms of cashews from Vietnam. All are part of the two-decade-long surge that has made imported food—often from developing countries—a significant component of the Canadian diet. All of it is grown or processed far beyond the reach of Canada's food inspection system.

In fact, domestic regulators are facing a new breed of competitor: private agribusiness. Consider the following examples: Wal-Mart (United States) has forced its hamburger meat suppliers to conduct specialized *E. coli* and salmonella tests; Loblaws has required suppliers for its private-label products to comply with Global Food Safety Initiative (a private retailer-created food safety organization) standards; and Costco forced its egg suppliers to vaccinate their hens against salmonella (Leeder, 2010). In each case, large food retailers have assumed the role of regulator usually carried out by national food safety authorities. Wal-Mart is instituting mandatory food inspections, Loblaws has set food safety regulations, and Costco is instituting preventative food safety requirements. In part, this trend in undoubtedly about preserving the company's brand name by, theoretically, minimizing the risk of a food crisis. After all, the risk of having the corporate name attached to a crisis is potentially ruinous—the Maple Leaf Foods listeriosis outbreak, for example.

While this movement is having an impact on the industry, its long-term impact is questionable. Critics claim that these efforts are less about safety and more about marketing. Additionally, if or when the potential return for not having these restrictions exceeds the potential cost of a food safety crisis, will the corporations quietly relax these standards? If so, does this "flexibility" demonstrate the fundamental need for a government-run food safety system? (Leeder, 2010)

As a systematic risk, the shifting boundaries of private and public sector actors represents an unknown. In fact, so too does this issue. Inherent to the food safety system is the knowledge that the safety of a product or food safety system can never be guaranteed, only shown to be false.

Risk Aversion

However, it is our natural instinct to fear for the worst—a reasonable concern given the terrible impact that these systematic risks present—that motivates much of our focus on food safety. After all, for every argument that food safety regulators are not doing enough to protect consumers there is the counterargument for evidence:

> The CFIA argues that the absence of big problems shows the system works. In any given day, [CFIA Chief Food Safety Officer] says, about 100-million meals are eaten in Canada—which works out to about 36.5 billion meals at year. And what's going wrong? There are about 250 to 300 recalls of food each year following inspections or consumer complaints. Canadians also suffer an estimated 11-million cases of acute gastroenteritis each year—a relatively minor amount—and one that federal authorities suggest is largely due to food preparation mistakes or bad hygiene.

Is this a concern for the agricultural system? After all, how can an abundance of caution about food safety—an absolutely critical issue—be misplaced? In fact, this can represent a concern when the public—an uninformed group of consumers—make decisions based on gut judgments. According to a November 2010 poll of Canadians, only 39.5% have more confidence in the current safety of food in Canada compared to 10 years ago (The Globe and Mail, 2010). This is, obviously, an opinion question to which there is no correct or incorrect answer. However, the dismal results suggest that there is notable apprehension about the effectiveness of the modern Canadian food safety system. This apprehension, if acted upon, can result in unnecessary, wrong-headed or frankly dangerous changes to Canada's food safety system.

To preserve Canada's record of safe food and excellent food safety systems—as illustrated by the 2010 *Food Safety Performance World Ranking Initiative* project—Canadians must take informed ownership of their food safety systems and make rational and educated decisions about our food safety regulators.

Anything else is too hard to swallow.

References

Chapter 1

Anonymous (2005) Cattle from Same Herd as Mad Cow May Have Made It to Food System: Officials. Moose Jaw Times Herald (January 8), p. 1.

Beachy, R. (2010) Science and sustainability: the emerging consensus. *Bioscience*, 60 (6), 406–407.

Bi, X., Wang, H. and Ge, J. (2010) The changes of densities and patterns of roads and rural buildings: a case study on Dongzhi Yuan of the Loess Plateau, China. *Environmental Monitoring and Assessment*, 164 (1–4), 549–560.

Brooymans, H. with files from Sean Gordon (2005) Mad Cow Food Fears: Meat from Infected Farm May Have Been Eaten. National Post (January 8), p. A1.

Collins, D. (2010) Heading for a world apocalypse? *The Journal of Social, Political, and Economic Studies*, 35 (2), 242–254.

Couture, J. (2009) Few Have Forgotten Recall: Survey. Leader Post (March 25), p. A.7.

Diekmeyer, P. (2008) Food flashpoint. *Canadian Grocer*, 122 (8), 49–51.

Dingman, J. (2008) Nanotechnology: its impact on food safety. *Journal of Environmental Health*, 70, 47–50.

Ferrari, A. (2010) Developments in the debate on nanoethics: traditional approaches and the need for new kinds of analysis. *NanoEthics*, 4 (1), 27–52.

Galloway, G. (2009) Company, Government Faulted in Listeria Deaths. The Globe and Mail (Index-only) (July 22), p. A.4.

Goveia, T. (2010) The maple leaf method. *Canadian Insurance*, 115 (4), S4.

Hughes, A., Wrigley, N. and Buttle, M. (2008) Global production networks, ethical campaigning, and the embeddedness of responsible governance. *Journal of Economic Geography*, 8 (3), 345–367.

Food Safety, Risk Intelligence and Benchmarking, First Edition. Sylvain Charlebois.

Kelleher, E. (2010) Bullish Food Forecasts Whet Investors' Appetite. Financial Times (July 17), p. 8.

Kilman, S. (2005) Alert in Latest Mad-Cow Case Was Delayed by a Misdiagnosis. Wall Street Journal (Eastern Edition) (June 27), p. A.2.

Labrecque, L., Charlebois, S. and Spiers, E. (2007) Can genetically modified foods be considered as a dominant design :an actor-network theory investigation of gene technology in agribusiness. *British Food Journal,* 109 (1), 81.

Lewis, R. and Tyshenko, M. (2009) The impact of social amplification and attenuation of risk and the public reaction to mad cow disease in Canada. *Risk Analysis,* 29 (5), 714–728.

Markovina, J. and Caputo, V. (2010) The impact of product designations on consumer decisions: the case of Croatian olive oil. *The Business Review, Cambridge,* 15 (1), 144–150.

Mason, C. (2009) Listeriosis probe identifies multiple deficiencies. *Canadian Medical Association Journal,* 181 (5), E88–E89.

Mehrjerdi, Y.R. (2010) Coupling RFID with supply chain to enhance productivity. *Business Strategy Series,* 11 (2), 107–123.

Moeller, J. (2010) Asia redraws the map of progress. *The Futurist,* 44 (5), 14–19.

Moore, K. (2010) Nanofertilizers, pee for your plants. *Chemical & Engineering News,* 88 (9), 88.

Morris, C. and Reed, M. (2007) From burgers to biodiversity? The McDonaldization of on-farm nature conservation in the UK. *Agriculture and Human Values,* 24 (2), 207–218.

Nicholson, S. (2009) Governing the gene: the politics of transgenic agriculture and the future of food. Ph.D. dissertation. The American University, United States—District of Columbia.

Nikiforuk, A. (2005). Beef Up the Science. The Globe and Mail (January 26), p. A.17.

Pechlaner, G. (2010) The sociology of agriculture in transition: the political economy of agriculture after biotechnology. *Canadian Journal of Sociology (Online),* 35 (2), 243–269.

Rosolen, D. (2010) Traceability solutions. *Food in Canada,* 70 (2), 37–38, 40.

Schade, C. and Pimentel, D. (2010) Population crash: prospects for famine in the twenty-first century. *Environment, Development and Sustainability,* 12 (2), 245–262.

Smith, J. (2010) Pasteurization Makes Meat Safer, Scientist Says Plants Taking Extra Step Require Fewer Inspections, Offer Consumers Better Protection, Professor Adds. Toronto Star (March 18), p. 8.

Smithers, J. and Joseph, A. (2010) The trouble with authenticity: separating ideology from practice at the farmers' market. *Agriculture and Human Values,* 27 (2), 239–247.

Spiekermann, U. (2009) Twentieth-century product innovations in the German food industry. *The Business History Review*, 83 (2), 291–315.
Zoellick, R. (2010) The end of the third world. *The International Economy*, 24 (2), 40–43.

Chapter 2

Agriculture and Agri-Food Canada (2002) Red Meat Industries Annual Report, Ottawa.
Agriculture and Agri-Food Canada (2009) *Exports—Agri-Food and Seafood for January to December 2008*, Agriculture and Agri-Food Canada, Ottawa, p. 1.
Agriculture and Agri-Food Canada (2010) *Canada at a Glance: Canadian Agri-Food Trade*, Agriculture and Agri-Food Canada, Ottawa, p. 1.
Bednarek, R. and Ahn, M. (2010) Trilogy natural products: managing a global distribution network for export growth1. *International Journal of Case Studies in Management (Online)*, 8 (2), 1–22.
Bello, W. (2008) How to manufacture a global food crisis. *Development*, 51 (4), 450–455.
Bow, B. (2010) Paradigms and paradoxes. *International Journal*, 65 (2), 371–380.
Buzby, J. (2003) International Trade and Food Safety: Economic Theory and Case Studies, USDA, Agricultural Economic Report No. (AER828) 145 pp.
Canadian Press (2003a) Le boeuf canadien se retrouve exclu des marches les plus importants. National News (May 21).
Canadian Press (2003b) Le plan d'aide pour le secteur bovin inclura des compensations et des prêts. National News (June 13).
Canadian Press (2003c) Jean Chrétien se fait rassurant au sujet du cas de la vache folle. National News (May 21).
Canadian Press (2003d) Anne McLellan affirme que les journalists alimentent la panique. National News (May 23).
Canadian Press (2003e) Les producteurs de boeuf canadiens ont besoin d'une aide urgente. National News (June 1).
Canadian Press (2003f) Les fermiers canadiens croient que les Américains agissent par vengeance. National News (June 15).
Charlebois, S. (2010) A Dash of Wisdom with Your Salt. Winnipeg Free Press (August 6), p. A14.
Couture, J. (2009) Few Have Forgotten Recall: Survey. Leader Post (March 25), p. A.7.
Cuddehe, M. (2009) Benefits Far Outshine Failings and Spats. Financial Times (November 10), p. 1.

Denis, H. (1993) *Comprendre à gérer les risques socio-technologiques majeurs*, Les Presses de l'Université de Montréal, UQTR, Chicago.

Duchesne, A. (2003) La vache folle sème l'inquiétude, il était une fois la vache folle.... La Presse (May 22), p. A3.

Eade, R. (2010) Sodium Content Starts to Come Down; Reducing Average Intake to 2,300 mg Still a Challenge. The Ottawa Citizen (August 5), p. F.2.

Frampton, C. (2008) TILMA: the impact of domestic trade pacts on learning environments. *Education Canada*, 48 (4), 66–69.

Ginsberg, R. (2007) Demystifying the European Union; enduring logic of regional integration. *Reference and Research Book News*, 22 (2), 154–162.

Gray, J. (2008) Up on the farm. *Canadian Business*, 81 (5), 23–24.

Hobbs, J. (2010) Public and private standards for food safety and quality: international trade implications. *The Estey Centre Journal of International Law and Trade Policy*, 11 (1), 136–152.

Jarrett, P. and Kobayakawa, S. (2008) Modernising Canada's agricultural policies. *OECD Economic Department Working Papers*, 1 (2), 5–30.

Kondro, W. (2010) Recalibration or "regurgitation"? *Canadian Medical Association Journal*, 182 (6), E257–E258.

Labrecque, J. and Charlebois, S. (2006) Conceptual links between two mad cow crises: the absence of paradigmatic change and policymaking implications. *International Food and Agribusiness Management Review*, 9 (2), 23–50.

Lang, T. and Heasman, M. (2004) *Food Wars, the Global Battle for Mouths Minds and Markets*, Earthscan, London.

Maniam, B., Hadley, L. and Thaler, R. (2003) North American free trade agreement-is it delivering what it promised? *Journal of American Academy of Business, Cambridge*, 2 (2), 379–385.

Martin, N. (2009) Schools Banned from Selling Food with Trans Fats. Winnipeg Free Press (September 10), p. B.3.

Pauchant, T. and Mitroff, I. (1992) *Transforming the Crisis-Prone Organization: Preventing Individual, Organizational, and Environmental Tragedies*, Jossey-Bass Publications, San Francisco, 256 pp.

Pearson, C. and Clair, J. (1998) Reframing crisis management. *The Academy of Management Review*, 23 (1), 59–76.

Phillips, P. (2001) Food safety, trade policy and international institutions, in *Governing Food: Science, Safety and Trade* (eds P. Phillips and R. Wolfe), McGill University Press/Queens School of Policy Studies, Montreal, pp. 27–48.

Sapp, S., Arnot, C., Fallon, J. *et al.* (2009) Consumer trust in the U.S. food system: an examination of the recreancy theorem. *Rural Sociology*, 74 (4), 525–545.

Schmidt, S. (2010) Meat Plant Okayed Before Outbreak; Overwork an Issue. Brandt Deficiencies Noted After Salmonella Cases. The Gazette (August 5), p. A.11.

Sweet, T., Balakrishnan, J., Robertson, B., Stolee, J. and Karim, S. (2010) Applying quality function deployment in food safety management. *British Food Journal*, 112 (6), 624–639.

Vogel, L. (2010) International food safety meeting sets melamine limits. *Canadian Medical Association Journal*, 182 (11), E521.

Weatherhill, S. (2009) *Report of the Independent Investigator into the 2008 Listeriosis Outbreak*, Government of Canada, Ottawa.

Chapter 3

Belgian Federal Agency for the Safety of the Food Chain (2009) *Fact and Figures: The Belgian Food Safety Agency in 2008*, Belgian Federal Agency for the Safety of the Food Chain, Brussels. pp. 9, 21, 27 and 33.

Canada National Microbiology Laboratory (2010) *National Enteric Surveillance Program*, http://www.nml-lnm.gc.ca/NESP-PNSME/index-eng.htm (accessed May 1).

Canadian Food Inspection Agency (2006) *Canadian Food Inspection Agency: Science and Regulation... Working Together for Canadians*, Canadian Food Inspection Agency, Ottawa, p. 4.

Canadian Food Inspection Agency (2010) *Guide to Food Labelling and Advertising*, Canadian Food Inspection Agency and Health Canada, Ottawa; Food Standards Australia New Zealand (2010) *Standard 1.2.2: Food Identification Requirements*, Canberra; Food Standards Australia New Zealand (2010) *Standard 1.2.5: Safe Marking of Food*, Canberra; Food Standards Australia New Zealand (2010) *Standard 1.2.8: Nutrition Information Requirements*, Canberra; Food Standards Australia New Zealand (2010) *Standard 1.2.4: Labelling of Ingredients*, Canberra; Food Standards Australia New Zealand (2010) *Standard 1.2.11: Country of Origin Requirements*, Canberra; and Food Standards Australia New Zealand (2010) *Standard 1.2.3: Mandatory Warning and Advisory Statements and Declarations*, Canberra.

Charlebois, S. and Yost, C. (2008) *Food Safety Performance World Ranking 2008*, Research Network in Food Systems, Regina, p. 27; Rocourt, J., Moy, G., Vierk, K. and Schlundt, J. (2003) *The Present State of Foodborne Disease in OECD Countries*, Food Safety Department, WHO, Geneva, p. 1.

Codex Alimentarius Commission (2008) *General Standard for the Labelling of Prepackaged Foods*, Joint FAO/WHO Food Standards Programme, FAO, Rome.

European Parliament (2007) *Directive 2000/13/EC of the European Parliament and of the Council of 20 March 2000 on the Approximation of the Laws of the Member States Relating to the Labelling, Presentation and Advertising of Foodstuffs*. OJ L 109, 6.5.2000. p. 29.

Ferri, M. (2005) Risk assessment and seafood products in developing countries, in *Report and Papers Presented at the FAO Workshop on Fish Technology, Utilization and Quality Assurance*, United Republic of Tanzania, Bagamoyo, p. 152.

Finland Ministry of Agriculture and Forestry (2006) *Food Safety in Finland*, Ministry of Agriculture and Forestry, Helsinki, pp. 2, 4, 5 and 7.

Food Safety Authority of Ireland (2009) *Annual Report 2008*, Food Safety Authority, Dublin, p. 8.

Galloway, G. (2010) Questions Raised About Gap in Food Inspection Standards. The Globe and Mail (March 15), p. A4.

Garrett, L. (2009) The path of a pandemic: how one virus spread from pigs and birds to humans around the globe. And why microbes like the H1N1 flu have become a growing threat. *Newsweek*, 153 (20), 943–951.

Genthner, F. (2010) *Safe Handling of Foods*, http://www.usdajapan.org/en/reports/Safe%20Handling%20of%20Foods.html (accessed May 1).

Japan Department of Food Safety (2010) *Inspection Results of Imported Foods Monitoring and Guidance Plan for FY 2009*, http://www.mhlw.go.jp/english/topics/importedfoods/09/09-04.html (accessed May 1).

Japan Ministry of Health, Labour and Welfare (2010) *Imported Foods Inspection Services Home Page*, http://www.mhlw.go.jp/english/topics/importedfoods/index.html (accessed May 1).

Kerr, W. (2009) Political precaution, pandemics and protectionism. *The Estey Centre Journal of International Law and Trade Policy*, 10 (2), 1–14.

Nierenberg, D. (2002) Food-borne Illness Widespread, in *Vital Signs 2002: The Trends That Are Shaping Our Future*, Worldwatch Institute, Washington, pp. 138–140.

Norway Lovdata (2010) *Regulations for Labeling of Foods*, http://translate.googleusercontent.com/translate_c?hl=en&ie=UTF-8&sl=auto&tl=en&u=http://www.lovdata.no/cgi-wift/ldles%3Fdoc%3D/sf/sf/sf-19931221-1385.html&prev=_t&rurl=translate.google.com&usg=ALkJrhhFRKBBOO1PNSzULZ26sSC14nFXEQ#map0 (accessed May 1).

Rocourt, J., Moy, G., Vierk, K. and Schlundt, J. (2003a) *The Present State of Foodborne Disease in OECD Countries*, Food Safety Department, WHO, Geneva, p. 4.

Rocourt, J., Moy, G., Vierk, K. and Schlundt, J. (2003b) *The Present State of Foodborne Disease in OECD Countries*, Food Safety Department, WHO, Geneva, p. 9.

Secretariat of the Joint FAO/WHO Food Standards Programme (2006a) *Understanding The Codex Alimentarius*, 3rd edn, Joint FAO/WHO Food Standards Programme, FAO, Rome, p. 19.

Secretariat of the Joint FAO/WHO Food Standards Programme (2006b) *Understanding The Codex Alimentarius*, 3rd edn, Joint FAO/WHO Food Standards Programme, FAO, Rome, p. ix.

Smith, J. (2010) Meat Plant Changed Some "Best Before" Dates: Report: Food Safety Concerns Raised About T.O. Plant. Toronto Star (June 30), p. A.12.

Switzerland Federal Office of Public Health (2009) *FOPH in Brief*, Federal Office of Public Health, Bern, p. 12.

Switzerland Federal Veterinary Office (2008) *Swiss Zoonoses Report 2008*, Swiss Government, Bern, p. 77.

United States Department of Agriculture (2010) *Food Safety and Inspection Service*, http://www.usda.gov/factbook/chapter9.htm (accessed May 1).

World Health Organization (2008) *WHO Initiative to Estimate the Global Burden of Foodborne Diseases: A Summary Document*, World Health Organization, Geneva, p. 2.

Chapter 4

Anonymous (2008) Food Safety Measures Ramped Up for Olympics. *Quality Progress*, 41 (6), 15.

Australia Department of Agriculture, Fisheries and Forestry (2009) *Review of Australia's Quarantine and Biosecurity Arrangements (the Beale Review)*, Department of Agriculture, Fisheries and Forestry, Canberra, pp. 4–5.

Australian Government (2010) *Counter-Terrorism White Paper: Securing Australia—Protecting Our Community*, Department of the Prime Minister and Cabinet, Canberra, p. 43.

Beale, R., Fairbrother, J., Inglis, A. and Trebeck, D. (2008) *One Biosecurity: A Working Partnership—The Independent Review of Australia's Quarantine and Biosecurity Arrangements Report to the Australian Government*, Attorney General's Department, Canberra, p. xiii.

Choe, Y., Park, J., Chung, M. and Moon, J. (2009) Effect of the food traceability system for building trust: price premium and buying behavior. *Information Systems Frontiers*, 11 (2), 167–179.

Cranfield, J., Henson, S. and Holliday, J. (2010) The motives, benefits, and problems of conversion to organic production. *Agriculture and Human Values*, 27 (3), 291–306.

Ecobichon, D.J. (2001) Pesticide use in developing countries. *Toxicology*, 160 (1), 27–33.

European Free Trade Area (2010) *The EFTA States*, http://www.efta.int/about-efta/the-efta-states.aspx (accessed May 1).

European Union European Environmental Agency (2007) *Europe's Environment: The Fourth Assessment*, European Environmental Agency, Copenhagen, p. 297.

Finland Ministry of the Interior (2010) *First National Counter-Terrorism Strategy Being Prepared*, Government Communications Unit, Helsinki.

Food Safety Network (2008) *Food Terrorism*, Food Safety Network, University of Guelph, Guelph, p. 2.

French Food Safety Authority (2004) *Évaluation des Risques Nutritionells et Sanitaires*, French Food Safety Authority, Paris, p. 18.

Government of Ireland, Expert Committee—Contingency Planning for Biological Threats (2002) *Biological Threats: A Health Response for Ireland*, Ministry for Health and Children, Dublin.

Halweil, B. (2004) Food Security Starts at Home. San Francisco Chronicle (December 23), p. B-7.

Hepner, I., Wilcock, A. and Aung, M. (2004) Auditing and continual improvement in the meat industry in Canada. *British Food Journal*, 106 (6/7), 553–568.

Hoffmann, V. (2009) What you don't know can hurt you: micronutrient content and fungal contamination of foods in developing countries. *Agricultural and Resource Economics Review*, 38 (2), 100–108.

Japan Food Safety Commission (2008) *Role of the Food Safety Commission*, Food Safety Commission Secretariat, Tokyo, p. 6.

Johnson, T. (2008) 10,000 Babies May Have Taken Contaminated Formula; Admission Casts Pall Over China's Dairy Industry. National Post (September 16), p. A.8.

Kathpal, T. and Kumari, B. (2009) Monitoring of pesticide residues in vegetarian diet. *Environmental Monitoring and Assessment*, 151 (1–4), 19–26.

Lizotte, R., Knight, S., Bryant, C. and Smith, S. (2009) Agricultural pesticides in Mississippi Delta oxbow lake sediments during autumn and their effects on *Hyalella azteca*. *Archives of Environmental Contamination and Toxicology*, 57 (3), 495–503.

Lucas, S. and Allen, P. (2009) Reducing the risk of pesticide exposure among children of agricultural workers: how nurse practitioners can address pesticide safety in the primary care setting. *Pediatric Nursing*, 35 (5), 308–317.

Macpherson, A. and McConnell, J. (2007) A survey of cross-border trade at a time of heightened security: the case of the Niagara bi-national region. *The American Review of Canadian Studies*, 37 (3), 301–321, 409.

McPeak, C., Devirian, J. and Seaman, S. (2010) Do environmentally friendly companies outperform the market? *Journal of Global Business Issues*, 4 (1), 61–66.

Milerski, B. (2010) Holocaust education in Polish public schools: between remembrance and civic education. *Prospects*, 40 (1), 115–132.
Mooney, P. and Hunt, S. (2009) Food security: the elaboration of contested claims to a consensus frame. *Rural Sociology*, 74 (4), 469–497.
Mustard, C., Bielecky, A., Etches, J. *et al.* (2010) Suicide mortality by occupation in Canada, 1991–2001. *Canadian Journal of Psychiatry*, 55 (6), 369–376.
Organisation for Economic Co-operation and Development. (2008) *Environmental Performance of Agriculture in OECD Countries Since 1990*. Paris, OECD. pp. 38, 39 and 44.
Peets, S., Gasparin, C., Blackburn, D. and Godwin, R. (2009) RFID tags for identifying and verifying agrochemicals in food traceability systems. *Precision Agriculture*, 10 (5), 382–394.
Public Health Agency of Canada (2010) *Bioterrorism and Emergency Preparedness*, http://www.phac-aspc.gc.ca/ep-mu/bioem-eng.php (accessed May 1).
Schoell, R. and Binder, C. (2009) System perspectives of experts and farmers regarding the role of livelihood assets in risk perception: results from the structured mental model approach. *Risk Analysis*, 29 (2), 205–222.
Starbird, A. and Amanor-Boadu, V. (2006) Do inspection and traceability provide incentives for food safety? *Journal of Agricultural and Resource Economics*, 31 (1), 14–26.
United Kingdom Centre for the Protection of National Infrastructure (2010) *Defending Food and Drink: Guidance for the Deterrence, Detection and Defeat of Ideologically Motivated and Other Forms of Malicious Attack on Food and Drink and Their Supply Arrangements*, British Standards Institution, London, p. iv.
United Kingdom Secretary of State for the Home Department (2010) *Pursue Prevent Protect Prepare: The United Kingdom's Strategy for Countering International Terrorism—Annual Report*, The Stationary Office, London, p. 16.
United States Department of Agriculture (2007) *Strategic Partnership Program Agroterrorism (SPPA) Frequently Asked Questions*, United States Department of Agriculture, Washington, p. 1.
United States Department of Agriculture (2010) *Strategic Partnership Program Agroterrorism Initiative*, United States Department of Agriculture, Washington, p. 1.
World Health Organization (2008) *Terrorist Threats to Food: Guidance for Establishing and Strengthening Prevention and Response Systems*, World Health Organization, Geneva, p. 1.
Yasuda, T. (2010) Food safety regulation in the United States: an empirical and theoretical examination. *The Independent Review*, 15 (2), 201–226.

Chapter 5

Canadian Food Inspection Agency (2010a) *Food Recall Fact Sheet,* http://www.inspection.gc.ca/english/corpaffr/recarapp/recafse.shtml (accessed May 1).

Canadian Food Inspection Agency (2010b) *2008–2009 Departmental Performance Report,* Canadian Food Inspection Agency, Ottawa.

Canadian Food Inspection Agency (2010c) *2008–2009 Estimates: Part III—Report on Plans and Priorities,* Canadian Food Inspection Agency, Ottawa, pp. 17, 22 and 26–43.

Canadian Food Inspection Agency (2010d) *Audit of the Management of Imported Food Safety,* Canadian Food Inspection Agency, Ottawa.

Cassels, S. (2006) Overweight in the Pacific: links between foreign dependence, global food trade and obesity in the Federated States of Micronesia. *Globalization and Health,* 2, 10.

Charlebois, S. and Yost, C. (2008) *Food Safety Performance World Ranking 2008,* Research Network in Food Systems, Regina, p. 49.

Crumb, M. (2010) Egg Recall Is Key Issue in Ag Secretary Race. The Associated Press (October 6).

European Commission, Directorate-General for Health and Consumer Protection (2009a) *The Rapid Alert System for Food and Feed (RASFF): Annual Report 2008,* European Commission, Directorate-General for Health and Consumer Protection, Luxembourg, p. 14.

European Commission, Directorate-General for Health and Consumer Protection (2009b) *The Rapid Alert System for Food and Feed (RASFF): Annual Report 2008,* European Commission, Directorate-General for Health and Consumer Protection, Luxembourg, p. 8.

European Parliament and European Council (2002) Regulation (EC) No. 178/2002 of the European Parliament and of the Council of 28 January 2002 Laying Down the General Principles and Requirements of Food Law, Establishing the European Food Safety Authority and Laying Down Procedures in Matters of Food Safety. Article 6, Section 3.

Food and Agriculture Organization of the United Nations and the World Health Organization (2004) *Second FAO/WHO Global Form of Food Safety Regulators—Proceedings of the Forum,* FAO, Rome, p. 3.

Food Standards Australia New Zealand. (2008) *Food Industry Recall Protocol: A Guide to Conducting a Food Recall and Writing a Food Recall Plan.* 6th. Food Standards Australia New Zealand: Canberra. pp. 4 and 11.

Food Standards Australia New Zealand (2009) *The Analysis of Food-Related Health Risks,* Food Standards Australia New Zealand, Canberra, p. 39.

Food Standards Australia New Zealand (2010) *Food Standards Australia New Zealand: Annual Report 2008–2009,* Food Standards Australia New Zealand, Canberra, p. 47.

Government of Canada (2004) Defining the responsibilities and tasks of different stakeholders within the framework of a national strategy for food control. Presented at the second FAO/WHO global forum of food safety regulators. Bangkok, Thailand, October 12–14.

Japan Director of the Department of Food Safety, Pharmaceutical and Food Safety Bureau (2010) *Development of Imported Foods Monitoring and Guidance Plan for FY 2010*, Ministry of Health, Labour and Welfare, Tokyo.

Japan Food Safety Commission (2010) *Role of the Food Safety Commission*, http://www.fsc.go.jp/english/aboutus/roleofthefoodsaftycommission.html (accessed May 1).

Japan Ministry of Agriculture, Forestry and Fisheries (2010) *General Provisions*, Japan Ministry of Health, Labour and Welfare, Tokyo Articles 1.3.2.5 and 1.3.2.6.

Japan Ministry of Health, Labour and Welfare (2009) *Health Risk Management System*, Japan Ministry of Health, Labour and Welfare, Tokyo, p. 1.

Japan Ministry of Health, Labour and Welfare (2010) *Imported Foods Inspection Services Home Page*, http://www.mhlw.go.jp/english/topics/importedfoods/index.html (accessed May 1).

Kolesnikov-Jessop, S. (2010) Singapore Looks to China for Food Security. The New York Times (September 27), pp. 833–839.

Norwegian Food Safety Authority (2014) *Reforming the Food Safety Administration in Norway*, Norwegian Food Safety Authority, Olso, p. 1.

Produce Safety Project (2010) *Legal and Regulatory Frameworks Governing the Growing, Packing and Handling of Fresh Produce in Countries Exporting to the U.S.*, The Pew Charitable Trusts at Georgetown University, Washington.

Sapp, S., Arnot, C., Fallon, J. *et al.* (2009) Consumer trust in the U.S. food system: an examination of the recreancy theorem. *Rural Sociology*, 74 (4), 525–545.

Switzerland Federal Office of Public Health (FOPH) (2010) *Negotiations Switzerland-EU on Food Safety and Public Health*, http://www.bag.admin.ch/themen/internationales/07419/07460/index.html?lang=en (accessed May 1).

United States Department of Agriculture (2010) *FSIS Recalls: Recall Case Archive*, http://www.fsis.usda.gov/fsis_recalls/Recall_Case_Archive_2009/index.asp (accessed May 1).

United States Department of Agriculture: Food Safety and Inspection Service (2008) *FSIS Directive: Recall of Meat and Poultry Products*, United States Department of Agriculture: Food Safety and Inspection Service, Washington, pp. 2, 6–14 and 16.

United States Food and Drug Administration (2010a) *2009 Recalls, Market Withdrawals, & Safety Alerts*, http://www.fda.gov/Safety/Recalls/ArchiveRecalls/2009/default.htm (accessed May 1).

United States Food and Drug Administration (2010b) *Regulatory Procedures Manual—Chapter 7: Recall Procedures*, United States Food and Drug Administration, Washington, pp. 1 and 2.
World Trade Organization (2010a) *Principles of the Trading System*, http://www.wto.org/english/thewto_e/whatis_e/tif_e/fact2_e.htm (accessed May 1).
World Trade Organization (2010b) *World Tariff Profiles 2009*, WTO Secretariat, Geneva, p. 34.

Chapter 6

Agriculture and Agri-Food Canada (2008) *Governments Announce Completion of the Growing Forward Multilateral Framework*, http://www.agr.gc.ca/cb/index_e.php?s1=n&s2=2008&page=n80711 (accessed May 1).
Alberta Agriculture and Rural Development (2010) *Growing Forward: Program Areas: Enhanced Food Safety: Traceability*, http://www.growingforward.alberta.ca/UCM01/ProgramAreas/EnhancedFoodSafety/index.htm (accessed May 1).
Albu, N. and Sterling, B. (2010) *Agriculture & Food Traceability Technology Vision*, OnTrace Agri-food Traceability, Guelph.
Buckley, M. and Reid, A. (2010) *Global Food Safety: Keeping Food Safe from Farm to Table*, American Society for Microbiology, Washington.
Canada Standing Committee on Agriculture and Agri-Food (2003) *The Investigation and the Government Response Following the Discovery of a Single Case of Bovine Spongiform Encephalopathy*, House of Commons Canada, Ottawa.
Canadian Food Inspection Agency (2003) *Summary of the Report of the Investigation of Bovine Spongiform Encephalopathy (BSE) in Alberta, Canada*, CFIA, Ottawa.
Canadian Food Inspection Agency (2010a) *Canada Advances System for Cattle Traceability*, http://www.inspection.gc.ca/english/corpaffr/newcom/2010/20100528e.shtml (accessed May 1).
Canadian Food Inspection Agency (2010b) *Traceability in Canada*, http://www.inspection.gc.ca/english/anima/trac/traccane.shtml (accessed May 1).
Codex Alimentarius Commission (2006) *Principles for Traceability/Product Tracing as a Tool Within a Food Inspection and Certification System*. CAC/GL 60-2006, Food and Agriculture Organization, Rome.
EFTA Surveillance Authority (2010a) *Food and Feed Safety*, http://www.eftasurv.int/internal-market-affairs/areas-of-competence/food-safety/food-and-feed-safety/ (accessed May 1).
EFTA Surveillance Authority (2010b) *The EFTA Surveillance Authority at a Glance*, http://www.eftasurv.int/about-the-authority/the-authority-at-a-glance-/ (accessed May 1).

European Commission, Directorate-General for Health and Consumer Protection (2007) *Food Traceability*. European Commission, Directorate-General for Health and Consumer Protection, Brussels, Belgium. p. 4.

European Parliament and European Council (2002) Regulation (EC) No. 178/2002 of the European Parliament and of the Council of 28 January 2002 Laying Down the General Principles and Requirements of Food Law, Establishing the European Food Safety Authority and Laying Down Procedures in Matters of Food Safety.

Golan, E., Krissoff, B., Kuchler, F. *et al.* (2004) Traceability in the U.S. Food Supply: Economic Theory and Industry Studies. United States Department of Agriculture: Economic Research Service. *Report No. 830.*

Goveia, T. (2010) The Maple leaf method. *Canadian Insurance*, 115 (4), S4.

Health Canada (2010) *Food Safety*, http://www.hc-sc.gc.ca/fn-an/securit/index-eng.php (accessed May 1).

Japan Ministry of Health, Labour and Welfare (2010) *Imported Foods Inspection Services Home Page*, http://www.mhlw.go.jp/english/topics/importedfoods/index.html (accessed May 1).

Japan Revision Committee on the Handbook for Introduction of Food Traceability Systems (2007) *Handbook for Introduction of Food Traceability Systems: Guidelines for Food Traceability*, Ministry of Agriculture, Forestry and Fisheries, Food Marketing Research & Information Center, Japan.

Kihm, U., Hueston, W. and Heim, D. (2003) *Report on Actions Taken By Canada in Response to the Confirmation of an Indigenous Case of BSE*, Canadian Food Inspection Agency, Bern.

Meat & Livestock Australia (2006a) *Australian Beef: Safe, Healthy and Delicious*, Meat & Livestock Australia, North Sydney.

Meat & Livestock Australia (2006b) *Australian Sheepmeat: Safe, Healthy and Delicious*, Meat & Livestock Australia, North Sydney.

Norway (2005) Act Relating to Food Production and Food Safety, etc. (Food Act).

Norway Ministry of Agriculture and Food (2010) *Specific Guidelines Regarding the Animal Welfare Act*, http://www.regjeringen.no/en/dep/lmd/whats-new/news/2009/mai-09/new-animal-welfare-act-/specific-guidelines-regarding-the-animal.html?id=562555 (accessed May 1).

Norway Universitetet I Oslo (2010) *Translated Norwegian Legislation*, http://www.ub.uio.no/cgi-bin/ujur/ulov/sok.cgi?type=FORSKRIFT (accessed May 1).

Norwegian Food Safety Authority (2010) *Import and Export* http://www.mattilsynet.no/english/import_export (accessed May 1).

OnTrace Agri-Food Traceability (2007) *Traceability Backgrounder*, OnTrace, Guelph.

Pauchant, T. and Mitroff, I. (1992) *Transforming the Crisis-Prone Organization: Preventing Individual, Organizational and Environmental Tragedies*, Jossey-Bass Publications, San Francisco.

Pearson, C. and Clair, J. (1998) Reframing crisis management. *The Academy of Management Review*, 23 (1), 59–76.
Prince Edward Island Department of Agriculture (2010) *Growing Forward: Traceability Projects Sub-Program*, http://www.gov.pe.ca/growingforward/index.php3?number=1025481&lang=E (accessed May 1).
Produce Marketing Association (2007) *Fresh Produce Imports into the US*, Produce Marketing Association, Newark.
Produce Safety Project (2010) *Legal and Regulatory Frameworks Governing the Growing, Packing and Handling of Fresh Produce in Countries Exporting to the U.S.*, The Pew Charitable Trusts at Georgetown University, Washington.
SafeMeat (2009) Annual Report 2007/08, Meat & Livestock Australia, North Sydney, p. 1.
SafeMeat (2010) *About Us*, http://www.safemeat.com.au/English/AboutUs.htm (accessed May 1).
United States Department of Agriculture: Animal and Plant Health Inspection Service (2010) *New Animal Disease Traceability Framework*, USDA: APHIS, Washington.
Weatherill, S. (2009) *Report of the Independent Investigator into the 2008 Listeriosis Outbreak*, Government of Canada, Ottawa.
World Health Organization (WHO) and Food and Agriculture Organization of the United Nations (FAO) (2006) *Understanding the Codex Alimentarius*, 3rd edn, Secretariat of the Joint FAO/WHO Food Standards Programme, Rome.

Chapter 7

AIG (2014) *Trends and Solutions in Combating Global Food Fraud*, www.aig.com/Jan-2014-NSF-Article_3171_567084.html (accessed September 26, 2014).
Alberta Health Services (2014) E. coli *Outbreak*, www.albertahealthservices.ca/10353.asp (accessed September 25, 2014).
Balazic, S., Wilcock, A., Hill, A. and Charlebois, S. (2013) Food safety performance: labeling and indications of allergens. *Food Protection Trends*, 33 (4), 232–239.
Blanquer, M., García-Alvarez, A., Ribas-Barba, L. *et al.* (2009) How to find information on national food and nutrient consumption surveys across Europe: systematic literature review and questionnaires to selected country experts are both good strategies. *British Journal of Nutrition*, 101 (Suppl. 2), S37–S50.
Business Development Bank of Canada (BDBC) (October 2013) *Mapping Your Future Growth: Five Game-Changing Consumer Trends.*

www.bdc.ca/Resources%20Manager/study_2013/consumer_trends_BDC_report.pdf (accessed September 26, 2014).

Canadian Food Inspection Agency (CFIA) (2012) *Independent Review of XL Foods Inc.* Beef Recall 2012, www.foodsafety.gc.ca/english/xl_reprt-rapprte.asp (accessed September 25, 2014).

Carpenter, J. and Tinker, R. (2012) Assessment of the Impact on Australia from the Fukushima Dai-ichi Nuclear Power Plant Accident. Australian Radiation Protection and Nuclear Safety Agency. *Technical Report Series No. 162.* www.arpansa.gov.au/pubs/technicalreports/tr162.pdf (accessed September 26, 2014).

Charlebois, S. and Hielm, S. (2014) Empowering the regulators in the development of national performance measurements in food safety. *British Food Journal*, 116 (2), 317–336.

Charlebois, S. and Watson, L. (2009) Risk communication and food recalls, in *The Crisis of Food Brands: Sustaining Safe, Innovative, and Competitive Food Supply* (eds A. Lindgreen, M.K. Hingley and J. Vanhamme), Gower, Burlington, pp. 29–44.

Charlebois, S. and Yost, C. (2008) *Food Safety Performance World Ranking: How Canada Is Doing. Research Network on Food Systems*, University of Regina, Regina.

Charlebois, S., von Massow, M. and Pinto, W. (2014a) Food recalls & risk perception: an exploratory case of the XL foods and the biggest food recall in Canadian history. *Journal of Food Products Marketing*, 20 (4), 1–17.

Charlebois, S., Sterling, B., Haratifar, S. and Naing, S.K. (2014b) Comparison of global food traceability regulations and requirements. *Comprehensive Reviews in Food Science and Food Safety*, 13 (5), 1104–1123.

Codex Alimentarius (1995) *Codex General Standard for Contaminants and Toxins in Food and Feed. Codex Standard 193-1995*, www.fao.org/fileadmin/user_upload/agns/pdf/CXS_193e.pdf (accessed September 26, 2014).

Corporate Research Associates Inc. (2012) *Food Safety: Canadians' Awareness, Attitudes and Behaviours (2011–12)*, http://epe.lac-bac.gc.ca/100/200/301/pwgsc-tpsgc/por-ef/canadian_food_inspection_agency/2012/029-11/report.pdf (accessed September 26, 2014).

FAOSTAT (2015) *Active Ingredient Use in Arable Land and Permanent Crops [t/1000 ha]. Pesticides, Agri-Environmental Indicators*, http://data.fao.org/ (accessed May 13, 2014).

Food Sentry (2014) *Analysis of International Food Safety Violations—2013*, http://www.foodsentry.org/analysis-international-food-safety-violations-2013/ July 31, 2014 (accessed August 13, 2014).

Food Standards Australia New Zealand (FSANZ) (2008) *Consumer Attitudes Survey 2007: A Benchmark Survey of Consumers' Attitudes to Food Issues*, Food Standards Australia New Zealand, Canberra.

Friends of Glass (2014) *Glass Health PR Survey Exploring Consumer Attitudes Related to Packaging and Food and Drink Safety*, http://vetropack.inettools.ch/upload/dokumente/europeanconsumersurvey_glass_health_fog.pdf (accessed June 2, 2014).

FSANZ (2014) *Food Recall Statistics*, www.foodstandards.gov.au/industry/foodrecalls/recallstats/Pages/default.aspx (accessed August 13, 2014).

García-Alvarez, A., Blanquer, M., Ribas-Barba, L. *et al.* (2009) How does the quality of surveys for nutrient intake adequacy assessment compare across Europe? A scoring system to rate the quality of data in such surveys. *British Journal of Nutrition*, 101 (Suppl. 2), S51–S63.

Health Canada (2000) *Canadian Guidelines for the Restriction of Radioactively Contaminated Food and Water Following a Nuclear Emergency*, http://www.hc-sc.gc.ca/ewh-semt/pubs/contaminants/emergency-urgence/index-eng.php (accessed May 21, 2014).

Health Canada (2012) *Food Allergies and Intolerances*, http://www.hc-sc.gc.ca/fn-an/securit/allerg/index-eng.php (accessed June 17, 2014).

Health Canada (2014) *Concentration of Contaminants & Other Chemicals in Food Composites*, http://www.hc-sc.gc.ca/fn-an/surveill/total-diet/concentration/index-eng.php (accessed May 30, 2014).

IFC (2008) *Food Safety Inspections: Lessons Learned From Other Countries*, International Finance Corporation, Washington.

International Food Information Council Foundation (IFICF) (2014) *More Americans Choosing Foods, Beverages Based on Healthfulness*, May 29, http://www.foodinsight.org/articles/more-americans-choosing-foods-beverages-based-healthfulness (accessed June 18, 2014).

Japan Ministry of Health, Labour and Welfare (JMHLW) (2008) *Food Poisoning Incidents and Cases, by Pathogenic Bacterium or Virus, 1991–2008*, http://idsc.nih.go.jp/iasr/31/359/graph/t3591.gif (accessed May 28, 2014).

Olsen, P. and Borit, M. (2013) How to define traceability. *Trends in Food Science and Technology*, 29 (2), 142–150.

Potter, A., Murray, J., Lawson, B. and Graham, S. (2012) Trends in product recalls within the agri-food industry: empirical evidence from the USA, UK and the Republic of Ireland. *Trends in Food Science and Technology*, 28 (2), 77–86.

Public Health Agency of Canada (PHAC) (2013a) *Public Health Notice:* E. coli *O157:H7 Illness Related to Cheese Produced by Gort's Gouda Cheese Farm*, November 15, http://www.phac-aspc.gc.ca/fs-sa/phn-asp/2013/ecoli-0913-eng.php (accessed July 28, 2014).

PHAC (2013b) *Public Health Notice:* E. coli *O157:H7 Illnesses in the Maritimes and Ontario*, February 7, http://www.phac-aspc.gc.ca/fs-sa/phn-asp/2013/ecoli-0113-eng.php (accessed July 28, 2014).

PHAC (2014a) *Notifiable Diseases On-Line*, http://dsol-smed.phac-aspc.gc.ca/dsol-smed/ndis/index-eng.php (accessed July 28, 2014).

PHAC (2014b) *National Enteric Surveillance Program*, https://www.nml-lnm.gc.ca/NESP-PNSME/index-eng.htm (accessed July 28, 2014).

SGS (2010) *Food for Thought: Study on Consumer Trust and Quality Awareness of Food*, September 23, http://www.sgs.com/en/Our-Company/News-and-Media-Center/News-and-Press-Releases/2010/09/Food-for-Thought-Study-on-Consumer-Trust-and-Quality-Awareness-of-Food.aspx (accessed July 21, 2014).

Soller, L., Ben-Shoshan, M., Harrington, D.W. *et al.* (2012) Overall prevalence of self-reported food allergy in Canada. *The Journal of Allergy and Clinical Immunology*, 130 (4), 986–988.

The Economist Intelligence Unit (2014) *Global Food Security Index*, http://foodsecurityindex.eiu.com/ (accessed September 26, 2014).

The World Bank (2014) *World Development Indicators. Population, Total*, http://data.worldbank.org/indicator/SP.POP.TOTL (accessed May 28, 2014).

Thomas, M.K., Murray, R., Flockhart, L. *et al.* (2013) Estimates of the burden of foodborne illness in Canada for 30 specified pathogens and unspecified agents, circa 2006. *Foodborne Pathogens and Disease*, 10 (7), 639–648.

UK Department for Environment, Food and Rural Affairs (2014) *Government Response to the Elliott Review of the Integrity and Assurance of Food Supply Networks*, September 2014, https://www.gov.uk/government/uploads/system/uploads/attachment_data/file/350735/elliott-review-gov-response-sept-2014.pdf (accessed October 1, 2014).

U.S. FDA (2005) *CPG Sec. 560.750 Radionuclides in Imported Foods: Levels of Concern. Inspections, Compliance, Enforcement, and Criminal Investigations*. Food and Drug Administration, http://www.fda.gov/ICECI/ComplianceManuals/CompliancePolicyGuidanceManual/ucm074576 (accessed May 22, 2014).

World Health Organization (WHO) (2013) *International Health Regulations (2005). IHR Core Capacity Monitoring Framework: Questionnaire for Monitoring Progress in the Implementation of IHR Core Capacities in States Parties. 2013 Questionnaire*, World Health Organization, Geneva.

WHO (2014) *Global Health Observatory Data Repository. 2012 and 2013 Data*, World Health Organization, Geneva. http://apps.who.int/gho/data/node.main.IHR11N?lang=en (accessed May 28, 2014).

Chapter 8

Agriculture and Agri-Food Canada (2010) *Agri-Food Regional Profile European Union: February 2006*, Agriculture and Agri-Food Canada, Ottawa.

Anonymous (2008) Food safety measured ramped up for Olympics. *Quality Progress*, 41 (6), 15.

Bambrough, K. (2009) *Sprott Resource Corp. Announced Launch of One Earth Farms Corp*, Sprott Resource Corp., Toronto.

Blay-Palmer, A. and Donald, B. (2006) A tale of three tomatoes: the new food economy in Toronto, Canada. *Economic Geography*, 82 (4), 383–399.

Bruwer, J. and Johnson, R. (2010) Place-based marketing and regional branding strategy perspectives in the California wine industry. *The Journal of Consumer Marketing*, 27 (1), 5–16.

Buckley, M. and Reid, A. (2010) *Global Food Safety: Keeping Food Safe from Farm to Table*, American Society for Microbiology, Washington.

Campbell, H. (2009) Breaking new ground in food regime theory: corporate environmentalism, ecological feedbacks and the 'food from somewhere' regime. *Agriculture and Human Values*, 26 (4), 309–319.

Center for Food Safety, et al. v. Thomas J. Vilsack, et al. (2010) United States District Court for the Northern District of California, C 08-00484 JSW.

Charlebois, S. and MacKay, G. (2010) Marketing culture through locally-grown products: the case of the Fransaskoisie *Terroir* products. *Problems and Perspectives in Management*, 8 (4), 90–102.

Choe, Y., Park, J., Chungand, M. and Moon, J. (2009) Effect of the food traceability system for building trust: price premium and buying behavior. *Information Systems Frontiers*, 11 (2), 167–179.

Codex Alimentarius Commission. (2006) *Principles for Traceability/ Product Tracing as a Tool Within a Food Inspection and Certification System*. CAC/GL 60-2006, Food and Agriculture Organization, Rome.

Deloitte (2010) *Deloitte 2010 Food Survey: Genetically Modified Foods*, Deloitte Development LLC, New York.

European Commission (2009) *Agricultural Trade Policy Analysis: Historical Price Volatility*, Directorate-General for Agriculture and Rural Development, Brussels.

European Parliament (2001) Directive 2001/18/EC of the European Parliament and of the Council of 12 March 2001 on the Deliberate Release into the Environment of Genetically Modified Organisms and Repealing Council Directive 90/220/EEC.

European Parliament (2003a) Regulation (EC) No. 1829/2003 of the European Parliament and of the Council of 22 September 2003 on Genetically Modified Food and Feed.

European Parliament (2003b) Regulation (EC) No. 1830/2003 of the European Parliament and of the Council of 22 September 2003 on Concerning the Traceability and Labelling of Genetically Modified Organisms and the Traceability of Food and Feed Products Produced from Genetically Modified Organisms and Amending Directive 2001/18/EC.

FAO (2009) Global agriculture towards 2050. How to Feed the World 2050: High-Level Expert Forum, Rome, October 12–13.
Gillam, C. (2010) Sugar Beet Ruling Pressures USDA GMO Oversight. Reuters (August 16), pp. 56–67.
Halweil, B. (2004) Food Security Starts at Home. San Francisco Chronicle (December 23), p. B-7.
Hamzaoui Essoussi, L. and Zahaf, M. (2009) Exploring the decision-making process of Canadian organic food consumers: motivations and trust issues. *Qualitative Market Research,* 12 (4), 443–459.
Health Canada (2010) *Genetically Modified (GM) Foods and Other Novel Foods,* http://www.hc-sc.gc.ca/fn-an/gmf-agm/index-eng.php (accessed May 1).
Johnson, T. (2008) 10,000 Babies May Have Taken Contaminated Formula; Admission Casts Pall Over China's Dairy Industry, National Post (September 16), p. A-8.
Leeder, J. (2010) Food Cop to Aisle Five: The Grocery Store's New Bag. The Globe and Mail (November 20), pp. 210–221.
Macpherson, A. and McConnell, J. (2007) A survey of cross-border trade at a time of heightened security: the case of the Niagara bi-national region. *The American Review of Canadian Studies,* 37 (3), 301–321.
McDonald, J. (2010) China Moves to Cool Inflation as Food Prices Surge. The Associated Press (November 17), pp. 101–112.
Monsanto Canada Inc. v. Schmeiser (2004) 1 S.C.R. 902, 2004 SCC 34.
Monsanto Co. v. Geertson Seed Farms (2010) United States Supreme Court, 130 S. Ct. 2743.
Moody, G. (2004) *Digital Code of Life: How Bioinformatics Is Revolutionizing Science, Medicine, and Business,* John Wiley & Sons, Inc., Hoboken.
Organisation for Economic Co-Operation and Development (2003) *Emerging Risks in the 21st Century: An Agenda for Action,* OECD, Paris.
Organisation for Economic Co-Operation and Development (2009) *The Bioeconomy Is Key to Tackling Many Future Global Challenges, Says OECD,* http://www.oecd.org/document/12/0,3343,en_2649_37437_42953484_1_1_1_1,00.html (accessed May 1).
Pew Research Center (2003) *Global Attitudes Project: Broad Opposition to Genetically Modified Foods,* http://pewglobal.org/2003/06/20/broad-opposition-to-genetically-modified-foods/ (accessed May 1).
Pollack, A. (2010) Judge Revokes Approval of Modified Sugar Beets. The New York Times (August 14), p. B1.
Produce Safety Project (2010) *Legal and Regulatory Frameworks Governing the Growing, Packing and Handling of Fresh Produce in Countries Exporting to the U.S,* The Pew Charitable Trusts at Georgetown University, Washington.

Saskatchewan Ministry of Agriculture (2010) *Saskatchewan Agri-Food Exports, 2003–2009,* Fact Sheet. Policy Branch, Regina.

Sims, R. (2009) Food, place and authenticity: local food and the sustainable tourism experience. *Journal of Sustainable Tourism,* 17 (3), 321–336.

Sprott Resource Corp (2010a) *Consolidated Financial Statements December 31, 2009 and 2008,* Sprott Resource Corp., Toronto.

Sprott Resource Corp (2010b) *Consolidated Financial Statements Third Quarter Ended September 30, 2010,* Sprott Resource Corp., Toronto.

The Globe and Mail (2010) Survey: Who Is Responsible for Keeping Our Food Safe? The Globe and Mail (November 19), pp. 923–927.

United Kingdom Centre for the Protection of National Infrastructure (2010) *Defending Food and Drink: Guidance for the Deterrence, Detection and Defeat of Ideologically Motivated and Other Forms of Malicious Attack on Food and Drink and Their Supply Arrangements,* British Standards Institution, London.

Ward, T. (2009) Reserve farming on the Canadian prairies 1870–1910. Canadian Network for Economic History Meeting: Halifax, September 23.

Wilhelmina, Q., Joost, J., George, E. and Guido, R. (2010) Globalization vs. localization: global food challenges and local solutions. *International Journal of Consumer Studies,* 34 (3), 357–366.

World Health Organization (2008) *Terrorist Threats to Food: Guidance for Establishing and Strengthening Prevention and Response Systems,* World Health Organization, Geneva.

World Health Organization (WHO) and Food and Agriculture Organization of the United Nations (FAO) (2006) *Understanding the Codex Alimentarius,* 3rd edn, Secretariat of the Joint FAO/WHO Food Standards Programme, Rome.

Index

Food Safety, Risk Intelligence and Benchmarking, First Edition. Sylvain Charlebois.